Energy, Climate and the Environment

Series Editor: **David Elliott**, Emeritus Professor of Technology Policy, Open University, UK

Titles include:

Marlyne Sahakian
KEEPING COOL IN SOUTHEAST ASIA
Energy Consumption and Urban Air-Conditioning

Benjamin K. Sovacool
ENERGY & ETHICS
Justice and the Global Energy Challenge

Joseph Szarka, Richard Cowell, Geraint Ellis, Peter A. Strachan and
Charles Warren (*editors*)
LEARNING FROM WIND POWER
Governance, Societal and Policy Perspectives on Sustainable Energy

Thijs Van de Graaf
THE POLITICS AND INSTITUTIONS OF GLOBAL ENERGY GOVERNANCE

Xu Yi-chong (*editor*)
NUCLEAR ENERGY DEVELOPMENT IN ASIA
Problems and Prospects

Energy, Climate and the Environment
Series Standing Order ISBN 978–0–230–00800–7 (hb)
978–0–230–22150–5 (pb)
(*outside North America only*)

You can receive future titles in this series as they are published by placing a standing order. Please contact your bookseller or, in case of difficulty, write to us at the address below with your name and address, the title of the series and the ISBNs quoted above.

Customer Services Department, Macmillan Distribution Ltd, Houndmills, Basingstoke, Hampshire RG21 6XS, UK.

The Growth of Biofuels in the 21st Century

Policy Drivers and Market Challenges

Robert Ackrill

Professor of European Economics and Policy, Nottingham Business School, Nottingham Trent University, UK

Adrian Kay

Professor of Government, Crawford School of Public Policy, Australian National University, Australia

First published 2014 by
PALGRAVE MACMILLAN

Palgrave Macmillan in the UK is an imprint of Macmillan Publishers Limited, registered in England, company number 785998, of Houndmills, Basingstoke, Hampshire RG21 6XS.

Palgrave Macmillan in the US is a division of St Martin's Press LLC, 175 Fifth Avenue, New York, NY 10010.

Palgrave Macmillan is the global academic imprint of the above companies and has companies and representatives throughout the world.

Palgrave® and Macmillan® are registered trademarks in the United States, the United Kingdom, Europe and other countries.

ISBN 978–1–137–30788–0

This book is printed on paper suitable for recycling and made from fully managed and sustained forest sources. Logging, pulping and manufacturing processes are expected to conform to the environmental regulations of the country of origin.

A catalogue record for this book is available from the British Library.

Library of Congress Cataloging-in-Publication Data
Ackrill, Robert, 1965–
    The growth of biofuels in the 21st century : policy drivers and market challenges/Robert Ackrill, Adrian Kay.
       pages   cm. — (Energy, climate and the environment)
    ISBN 978–1–137–30788–0 (hardback) — ISBN 978–1–137–30788–0
    1. Biomass energy—Government policy.   2. Renewable energy sources—Government policy.   3. Energy policy—Environmental aspects.   4. Energy resources development.   5. Biomass energy industries.   I. Kay, Adrian.   II. Title.
    HD9502.5.B542A35 2014
    333.95'39—dc23   2014028113

Contents

Tables, Figures and Boxes

Tables

Figures

Boxes

Series Editor's Preface

Concerns about the potential environmental, social and economic impacts of climate change have led to a major international debate over what could and should be done to reduce emissions of greenhouse gases. There is still a scientific debate over the likely *scale* of climate change, and the complex interactions between human activities and climate systems, but global average temperatures have risen and the cause is almost certainly the observed build-up of atmospheric greenhouse gases.

Whatever we now do, there will have to be a lot of social and economic adaptation to climate change – preparing for increased flooding and other climate-related problems. However, the more fundamental response is to try to reduce or avoid the human activities that are causing climate change. That means, primarily, trying to reduce or eliminate emission of greenhouse gases from the combustion of fossil fuels. Given that around 80 per cent of the energy used in the world at present comes from these sources, this will be a major technological, economic and political undertaking. It will involve reducing demand for energy (via lifestyle choice changes – and policies enabling such choices to be made), producing and using whatever energy we still need more efficiently (getting more from less) and supplying the reduced amount of energy from non-fossil sources (basically switching over to renewables and/or nuclear power).

Each of these options opens up a range of social, economic and environmental issues. Industrial society and modern consumer cultures have been based on the ever-expanding use of fossil fuels, so the changes required will inevitably be challenging. Perhaps equally inevitable are disagreements and conflicts over the merits and demerits of the various options and in relation to strategies and policies for pursuing them. These conflicts and associated debates sometimes concern technical issues, but there are usually also underlying political and ideological commitments and agendas which shape, or at least colour, the ostensibly technical debates. In particular, at times, technical assertions can be used to buttress specific policy frameworks in ways which subsequently prove to be flawed.

The aim of this series is to provide texts which lay out the technical, environmental and political issues relating to the various proposed policies for responding to climate change. The focus is not primarily on the viii *Series Editor's Preface*

science of climate change, or on the technological detail, although there will be accounts of the state of the art, to aid assessment of the viability of the various options. However, the main focus is the policy conflicts over which strategy to pursue. The series adopts a critical approach and attempts to identify flaws in emerging policies, propositions and assertions. In particular, it seeks to illuminate counter-intuitive assessments, conclusions and new perspectives. The aim is not simply to map the debates, but to explore their structure, their underlying assumptions and their limitations. Texts are incisive and authoritative sources of critical analysis and commentary, indicating clearly the divergent views that have emerged and also identifying the shortcomings of these views.

That certainly applies to the present text, which explores the biofuels policy debate. It takes the adage that 'there is no such thing as good biofuels and bad biofuels, only biofuels done well and biofuels done badly' as a guide and tries to make sense of the often very charged debate over food v. fuel and land-use changes. The authors attempt to be neutral, but it is clear they are unhappy with the way the debate has been pursued, and they identify what they see as misconceptions in relation to some of the food/land-use issues. Drawing on material from Brazil, the EU and US, they offer a useful and critical addition to the extensive academic and NGO literature on biofuel policy issues.

Preface and Acknowledgements

This is a book about policies. Specifically, it is a book that explores why governments have devoted so much attention and enormous intellectual, political and financial resources in recent years to promoting the production of biofuels, and their use in road transport. We look at both how governments have done this, and why. Whilst governments around the world are promoting biofuels, we focus on the three cases that dominate the production and use of biofuels – Brazil, the EU and US. We also focus on the two biofuels which have come to dominate not only these policies, but also the debates circling around biofuels – ethanol and biodiesel.

In addition to their market dominance, these three cases also offer important points of policy comparison and contrast. The EU and US have promoted biofuels on a significant scale only since the millennium, a development which provides us with the title of this book. As well as significant similarities between their policies and their experiences of biofuels promotion, we shall see important differences in their domestic context, design and consequences. Brazil, meanwhile, offers us two distinct policies. One, for ethanol, stands in sharp contrast to recent developments in the EU and US; the other, promoting biodiesel, offers elements of similarity as well as difference.

One feature all of these policies share, however – and this has really struck us as we have written up our research for this book – is the sheer scale of ambition policy-makers had in embarking on the path to the creation of biofuels markets. The goal of creating, in many cases almost from scratch, and frequently in only a few years, both the supply of, and demand for, biofuels, is not only massively ambitious; it also appears to fly in the face of decades of experience marked by the progressive withdrawal of the state from markets. For reasons elaborated on later, we cannot judge the wisdom of this decision in definitive and absolute terms of policy success or failure. We can, however, say that for reasons expanded on later, the scale of this policy ambition has, of itself, created challenges that policy-makers are still grappling with. Biofuels policies have, unambiguously, driven a massive expansion in biofuels production and use in recent years – but analysis of this evolving work-in-progress requires looking at more than simply the quantity of biofuels now out there.

Since the turn of the millennium, this dramatic increase in the volume of biofuels production and use has generated a similar increase in the debates and rhetoric surrounding biofuels. Views range from those expressed in 1925 by Henry Ford, who described ethanol as 'the fuel of the future', to biofuels being described in 2007 by Jean Ziegler (the United Nations special rapporteur on the right to food) as 'a crime against humanity'. We offer this book neither as an apologia for biofuels, nor as a medium for the blanket condemnation of biofuels. Instead, we seek to take a viewpoint that, as far as such a controversial subject will allow, it is an objective and neutral stance on biofuels. As such, we are aware that we are quite likely to be seen by some as offering a pro-biofuels analysis, simply because we are not specifically anti-biofuels.

The essence of our position can be summed up neatly with reference to a statement heard often in Brazil: 'there is no such thing as good biofuels and bad biofuels, only biofuels done well and biofuels done badly.' What this says, first, is that there is nothing intrinsic in biofuels that renders them either good or bad, a statement predicated on the multiplicity of ways in which different biofuels can be produced. This, however, leads into the second point that there are many different ways of producing biofuels, and these *can* be located along a spectrum of good to bad, depending on the specific details of a particular production method and its consequences.

A theme running through this book is therefore what distinguishes biofuels done well from biofuels done badly, and, linked to this, where uncertainty and ambiguity arise around biofuels and their policies, as a result. We therefore aim to inform the reader about the policies being used in Brazil, the EU and US, and how these policies are being influenced and challenged by questions of how biofuels are being 'done'. In this way, we aim to give the reader a better understanding, an informal toolkit if you will, to help make more informed judgements about biofuels and the biofuels debate. To provide a clear structure to our analysis, this book is split into two distinct parts.

Part I introduces biofuels and biofuels policies. We refer to the chemistry, science and processes of making biofuels only insofar as it is necessary to our introduction to biofuels, presented in Chapter 1. We also identify the key themes surrounding biofuels policies that will provide the analytical bedrock for the rest of the book. In Chapters 2–4, we discuss key policy developments in our three focus countries, locating them in their historical context (we use the word 'country' as convenient shorthand, recognising of course that the EU is not a country, but an international organisation made up of, currently, 28 countries). Chapter 5 then summarises the main findings of Chapters 2–4, with the analysis structured around the key themes identified in Chapter 1. In Chapter 5, we identify key similarities and differences between our three policy cases in terms, principally, of the three drivers of biofuels policies – energy security, climate change mitigation and rural development.

Part II of the book offers a more formal analysis of some of the key issues and challenges around biofuels policy-making. Chapter 6 provides an overview of the challenges of policy design and policy implementation. Chapters 7 and 8 analyse the international and global aspects of biofuels policy-making. Chapter 7 explores the specific issue of how domestic policy, in implementation, has thus far resisted a number of significant external pressures, whilst Chapter 8 investigates external dimensions of the domestic policy, notably, how the trade rules and obligations of the World Trade Organization (WTO) have influenced domestic policy design, and how economic globalisation creates implementation challenges. The biofuels case offers a fascinating insight into a broader category of policy challenges: how governments attempt to govern in an increasingly globalised market economy. The answer, as provided by biofuels policies, is that they can do so only in conjunction with (multinational) firms, non-governmental organisations and international organisations such as the WTO.

Because biofuels policies face challenges now, and will continue to do so as production and use expand in the future, Chapter 9 analyses some of these challenges. We conclude, in Chapter 9, where we began in Chapter 1: whilst biofuels and biofuels policies have extremely important technical dimensions, in this book we view biofuels policies, the challenges they seek to address and the challenges they can give rise to, as an essentially political problem. In particular, these policies are seeking to reconcile diverse interests and incommensurable values. There are challenging questions over, for example, the debates surrounding food v. fuel and indirect land-use change. Often, these have been used in a partial and tendentious fashion by vested interests, for and against biofuels and, within this, for different types of biofuel and biofuel policy. The contested politics of biofuels policies endures; and whether policy-makers possess the capacity to deal with these challenges is not clear – as we shall see.

The research for this book has had a long gestation. In 2007, The Leverhulme Trust awarded Rob Ackrill a Research Fellowship (RF/7/RFG/2007/0152) to support research into the factors behind the

2005 reform of the EU sugar policy. Part of the time on the project was spent in Brisbane, where he and Adrian Kay (who, at the time, was at Griffith University) very quickly realised that sugar was opening up several potential avenues of investigation. We therefore decided to park biofuels policy and focus just on sugar. We came back to biofuels in 2008 and, in 2009, were awarded a Small Grant by the Economic and Social Research Council (RES-000-22-3607). This book is a product of that research – and it is this research project which defines the boundaries of its content (see also Ackrill and Kay, 2012, for a concise summary of some of the main themes and findings). We are extremely grateful, both to the Leverhulme Trust for setting the ball rolling and for the ESRC for funding this subsequent research project.

Their support has allowed us to travel to Brussels, Washington DC and Brasilia, to conduct interviews, host research seminars and otherwise meet and engage with over 50 senior individuals involved in biofuels. In Brussels, these included officials in several of the Directorates-General central to biofuels policy, to MEPs and to people working in a range of sectoral organisations with an interest in EU biofuels policy. In Washington, interviews were held with officials in several of the key Departments and Agencies, staffers working with key committees in both Houses of Congress and with people working in a range of sectoral organisations with an interest in US biofuels policy. In Brasilia, most interviews were conducted with officials in the key ministries involved in Brazilian biofuels policy, and also with academics.

Given the highly sensitive nature of the policy work on biofuels, it was a condition of access to more or less every interviewee and participant that we preserve anonymity. Thus, in this book, when referencing information obtained in a particular interview, we give some indication of the position held by a particular interviewee – and we can assure the reader that all interviewees were in senior and influential positions in policy circles – but we cannot give such information as would allow for an individual's identity to be inferred from what we write. We are indebted to all interviewees and seminar participants, for giving so willingly of their time and knowledge.

A first round of interviews was conducted in Washington, Brasilia and Brussels during 2010. We then returned in 2011, where we presented our initial findings. These meetings – a combination of larger formal seminars smaller, more informal, individual meetings – provided us with feedback on our work to date and a wealth of new material to absorb. We were fortunate to be able to present formal seminars at the

European Commission (Energy Directorate General), at the US Department of Agriculture and US Department of Energy, as well as have more informal meetings at the US Environmental Protection Agency, US State Department; and, in Brazil, at the Ministry of External Relations and the Ministry of Mines and Energy.

In addition to the primary interview data collected for this research, we have also made extensive use of the large and rapidly expanding body of available research on biofuels and biofuels policies. As well as giving this book a wider analytical base than could be provided just from our own data collection, this also serves the important function of allowing us to triangulate our interview data with other, published, materials. Given that we have written this book for a wider target audience than just academics, we have tried to accommodate the issue of accessibility of materials, given that many academic journals offer subscription-only access. Where possible, we have included materials freely available via the Internet. In addition, we wish to point out for UK readers, a new initiative called 'Access to Research', being piloted from early 2014. This will see some academic publications being made available through public libraries. See: http://www.accesstoresearch.org.uk/

As well as interviews and meetings with policy-makers, officials and others with a professional interest in biofuels, we have given many presentations to academic audiences. The opportunity to participate in these meetings, and to receive comments and feedback from colleagues, has helped us considerably in developing further the ideas set out in this book. As part of the second round of interviews and visits in 2011, we gave seminars at The World Bank in Washington DC; at Miami-Florida European Union Center of Excellence, Florida International University Miami; and the Centre for Sustainable Development, University of Brasilia. Since the start of the project, we have also delivered papers at numerous academic seminars, workshops and conferences across Europe, and in the US, Singapore, New Zealand and Australia.

Rob Ackrill is extremely grateful to Jacqueline Lo, Jane Coultas and everyone at the Centre for European Studies, Australian National University, Canberra, for their warm and generous hospitality. The award of Visiting Fellowships in July–August 2012 and April–May 2014 provided an oasis where work could be undertaken in the crucial initial and final phases of writing for this book. We are both grateful to the Centre for their hosting a Roundtable on the Development and Impact of Biofuel Policies, held on 3 August 2012 under Chatham House Rules.

Particular thanks owed to David Elliott, the Series Editor, to Ambra Finotello, our Editorial Assistant who did a great deal of assisting, and to

everyone at Palgrave who has helped and supported us as we have bat-
tled with the writing of this book, despite having their patience tested to
the limit as we have struggled (and failed) to maintain a balance between
the differing demands of academic life whilst trying to sustain progress
on this manuscript.

Finally, we offer special thanks to Ursula and Siwan for all their love
and support. They provided two essential ingredients for the book:
empathetic knowledge of when not to ask how it was going, combined
with a steadfast commitment not to take too much of an interest in
biofuels.

Abbreviations

AB	Appellate Body (WTO)
AoA	Agreement on Agriculture (WTO)
API	American Petroleum Institute
B2/B5/B10	diesel blended with the quoted percentage of biodiesel
CAFE	Corporate Average Fuel Economy
CAP	Common Agricultural Policy (EU)
CARB	California Air Resources Board
CCS	Carbon Capture and Storage
CIMA	Conselho Interministerial do Açúcar e do Álcool (Interministerial Council of Sugar and Alcohol, Brazil)
CNPE	Conselho Nacional de Politica Energética (National Council for Energy Policy, Brazil)
COP	Conference of the Parties (to the UNFCCC, defined below)
DDGS	Dried Distillers Grains with Solubles
DG	Directorate General (of the European Commission)
DG-ENER	Energy Directorate General of the European Commission
DGS	Distillers Grains with Solubles
DLUC	Direct Land-Use Change
DoE	US Department of Energy
DSP	Dispute Settlement Procedure of the World Trade Organisation
E10/E15/E25/E85	petrol blended with the quoted percentage of ethanol
EISA	Energy Independence and Security Act of 2007 (US)
EPA	Environmental Protection Agency (US)
ETS	Emissions Trading Scheme (EU)
FAO	Food and Agriculture Organization (UN)
FFVs	flex-fuel vehicles
FQD	Fuel Quality Directive of 2009 (EU, revised)
GATT	General Agreement on Tariffs and Trade

GBEP	Global Bioenergy Partnership
GHGs	greenhouse gases
GMA	Grocery Manufacturers Association (US)
GMO	Genetically modified organism
GREET	Greenhouse gases, Regulated Emissions, and Energy use in Transportation
HS	Harmonised Commodity Description and Coding System
IEA	International Energy Agency
IFPRI	International Food Policy Research Institute
IIT	Intra-Industry Trade
ILUC	Indirect Land-Use Change
IMF	International Monetary Fund
IPCC	Intergovernmental Panel on Climate Change
ISO	International Standards Organization
LUC	Land-Use Change
MAPA	Ministério da Agricultura, Pecuária e Abastecimento (Ministry of Agriculture, Livestock and Supply, Brazil)
MEP	Member of the European Parliament
MFN	Most-Favoured Nation
MME	Ministério de Minas e Energia (Ministry of Mines and Energy, Brazil)
MNC	Multinational Company (or Corporation)
MTBE	Methyl Tertiary Butyl Ether
NGO	Non-Governmental Organisation
OECD	Organisation for Economic Cooperation and Development
OPEC	Organisation of the Petroleum Exporting Countries
PNPB	Programa Nacional de Produção e Uso de Biodiesel (National Program for the Production and Use of Biodiesel, Brazil)
PPM	Processing and Production Method
RED	Renewable Energy Directive of 2009 (EU)
RFS	Renewable Fuel Standard (US biofuels policy)
RIN	Renewable Identification Number
RSB	Roundtable on Sustainable Biomaterials (previously the Roundtable on Sustainable Biofuels)
RSPO	Roundtable on Sustainable Palm Oil
RTRS	Roundtable on Responsible Soy
SCM	Subsidies and Countervailing Measures agreement (WTO)
SEA	Single European Act
SEM	Single European Market
SFC	Social Fuel Certificate

TBTA	Technical Barriers to Trade Agreement (WTO)
UNFCCC	United Nations Framework Convention on Climate Change
UNICA	União da Indústria de Cana-de-Açúcar (Brazilian Sugarcane Industry Association)
USDA	United States Department of Agriculture
VEETC	Volumetric Ethanol Excise Tax Credit (US)
WTO	World Trade Organization

Part I

1
Biofuels and Biofuels Policies – An Introduction

Introduction

In recent years, for a number of reasons, governments have been increasing their efforts considerably to promote the production and use of energy derived from renewable sources. Concerns have been raised, for example, over the dependence on fossil fuels, the utilisation of which has considerable climate and environmental impacts. Moreover, in the case of oil, this creates an economic dependency of the vast majority of countries globally which lack oil resources upon the limited number of countries which have those resources, with many of whom political relations, and thus trade, are seen to be unstable and unreliable. There is also a widely held economic concern with oil that we are seeing the depletion of finite reserves, with consequences also for the price of oil.[1]

Renewable energies offer the chance to alleviate all of these concerns, whether they are directed at power (electricity) generation, heating (and cooling/air conditioning) and (road) transport fuels. That said, it cannot be taken for granted that just because an energy source is renewable and aids a shift away from fossil-fuel dependence, it is in all ways superior to the fossil fuel replaced. This is particularly the case with biofuels, the dominant form of renewable energy used in the transport fuel mix. Indeed, an analysis of the unintended consequences and side effects of biofuels policies, production and use represents a major theme of this book.

Oil, as noted, has a number of problems associated with it. Yet whilst the global energy matrix has, in the 40 years since the first oil crisis of 1973/74, reduced its dependence on oil considerably, transport fuel remains almost 100 per cent oil-dependent.[2] This goes some way to explaining the 'biofuels frenzy'[3] seen since the turn of the millennium – and also the relative lack of attention paid initially to the potential downsides of different types of biofuel. This is, however, only one dimension driving the biofuels frenzy. Later in this chapter, we explore this and the other key policy drivers more fully.

This book is written neither an apologia for biofuels, nor an assault on them. Our intention, in contrast to almost all of the growing number of books on this subject, is to remain neutral in this debate. Rather, we take as our starting point a saying that we heard repeatedly in Brazil: there is no such thing as good biofuels or bad biofuels; only biofuels done well and biofuels done badly. There is nothing intrinsically good or bad about using renewable fuel in the transport fleet. Rather, what matters in determining whether biofuels are done well or done badly is a range of complex factors, where different types of biofuel policy promote different biofuels, produced in many different ways, derived from many different types of feedstock, with their performance judged against multiple criteria.

Our central aims, by the end of the book, are to analyse criteria by which biofuels can be judged as having been done well or badly, examine the trajectory of biofuels-promoting policies and analyse the links between policies and the delivery (or not) of the features policy-makers desire of biofuels. In so doing, we wish to allow the reader to be better-informed about the range of factors which need to be considered in order to make his or her own mind up about different types of biofuel and biofuel policy, the range of possible policies and policy outcomes that are possible, and the challenges policy-makers face when designing and reforming policies.

Our particular focus is on the following aspects: the policies of Brazil, the EU and US as the three dominant players in global biofuels markets (in 2012, Brazil, the EU and US produced over 90 per cent of global ethanol, and nearly 80 per cent of global biodiesel); ethanol and biodiesel as the two dominant forms of biofuel (very nearly 100 per cent of the market to date); and land transport as the consumer of biofuels. There are renewable transport fuels other than ethanol and biodiesel, being developed for aviation as well as road transport, but they are still essentially in the development stage, with almost no market penetration thus far. It is beyond the scope of this book to explore these, but the interested reader can follow this up via several of the references cited in this chapter.

What are biofuels?[4]

This is, primarily, a book about policies, policies used by governments around the world – but especially in our three focus countries – to promote the production and use of biofuels. In the EU and US, more-over, these policies are seeking to promote a large expansion in biofuels production and use, from a low base, in a relatively very short period of time. In this section, we begin with an introduction to biofuels them-selves, recognising that whilst an elementary understanding of some of the technical aspects of biofuels is helpful to understanding the policy story we tell later, a detailed knowledge of the science of biofuels is not.[5]

At their most basic level, biofuels are fuels extracted or fermented from organic matter. Whilst there are several types of biofuel available for use as transport fuel, we focus on the two which dominate biofuels markets and the attention of policy-makers: ethanol and biodiesel. Moreover, each can be made from a range of inputs/feedstocks which can be classified in a number of ways, each emphasising different sets of characteristics. We look at the ethanol–biodiesel distinction next, followed by an analysis of how different biofuels (both ethanol and biodiesel) can be classified, based on the feedstocks from which they are derived.

Ethanol and biodiesel

Ethanol is derived from sugars, whilst biodiesel is derived from oils. Ethanol production is dominated by the US and Brazil which use, respec-tively, corn and sugarcane. With the former, sugars are extracted from the corn starch, whereas with sugarcane, there is a 'direct' route to the ethanol, via fermentation. One of the shortcomings of first gen-eration ethanol derived from starch is that the starch itself represents a relatively small percentage of the total volume of biomass presented for processing. One potential benefit arising from the greater com-mercial development of second and third generation ethanol processes (discussed in the next subsection) could be to allow for the greater con-version of more of the total volume of biomass processed. This should mean, for example, a higher volume of biofuel produced per unit weight of biomass – which, depending on the feedstock, could also result in the delivery of a higher volume of biofuel per unit area of land used to grow the feedstock. Even so, this latter benefit would potentially be offset by the fact that some of these feedstocks would still involve the utilisation of land, and of a crop that can be used as food (we return to these issues in Chapter 9).

Ethanol is, typically, blended with petrol. Ethanol has a lower energy content per unit volume than petrol (roughly 70 per cent) but, with a higher octane level, it also improves the performance of the petrol. By improving the combustion of the fuel, it helps lower a range of emis-sions, such as carbon monoxide and sulphur oxide, as well as a variety of carcinogens. It may, however, lead to a slight elevation in the level of nitrogen oxide in the air. Engines have been developed which can run on very high ethanol blends, or even pure ethanol. These flex-fuel engines are utilised extensively in Brazil (see Chapter 2), but sales of flex-fuel vehicles (FFVs) are also growing in the US, especially the Midwest corn belt where E85, fuel blended to 85 per cent ethanol, is widely available. They are also utilised in parts of the EU, for example, Sweden. There remains no agreement over the level to which ethanol can be blended into petrol before engines need modification to avoid damage. Imported non-FFV vehicles in Brazil use domestic petrol which comes, typically, as a 25 per cent ethanol blend; whilst in the US, approval has only recently been given for E15 – and then only for newer vehicles. Interviews revealed that, in the EU, carmakers were reluctant to issue warranty cover even for E10.

Biodiesel production is slightly less concentrated globally than ethanol, but is still dominated by the EU, US and Brazil. The choice of (first generation) feedstocks is also more varied, and includes soybeans in the US; soybeans, castor and palm in Brazil; rapeseed[6] in the EU; and oil palm in several tropical-belt countries. The two dominant forms of conversion of the feedstock are transesterification and hydrogenation, of which the former is the more widely utilised process (IEA Bioenergy, 2009: 34).[7] According to the Food and Agriculture Organisation of the United Nations (FAO, 2008: 13), the performance gap between diesel and biodiesel is closer than between ethanol and petrol, with biodiesel having 88–95 per cent of the energy content of diesel. Moreover, biodiesel shares many of the engine functionality and emissions advantages of ethanol.

Despite these advantages, the production of biodiesel continues to lag behind ethanol. One likely reason for this is that, of the major biofuels players, diesel is an important fuel for cars and light vehicles only in the EU. Compared with ethanol, in both Brazil and the US biodiesel is a much smaller – but nonetheless growing – part of the biofuels scene, as Chapters 2 and 4 elaborate. Indeed, interviews in Brasilia with officials who are involved in Brazil's biodiesel policy suggested that economic growth was expected to result in the demand for diesel doubling in ten years. This is because the country's infrastructure means that the movement of people and goods is dominated by road transportation: buses, coaches and lorries.

Figures 1.1 and 1.2 show the production levels and shares of our three focus countries, respectively, for ethanol and biodiesel. As already indicated, the three countries dominate world ethanol production, with

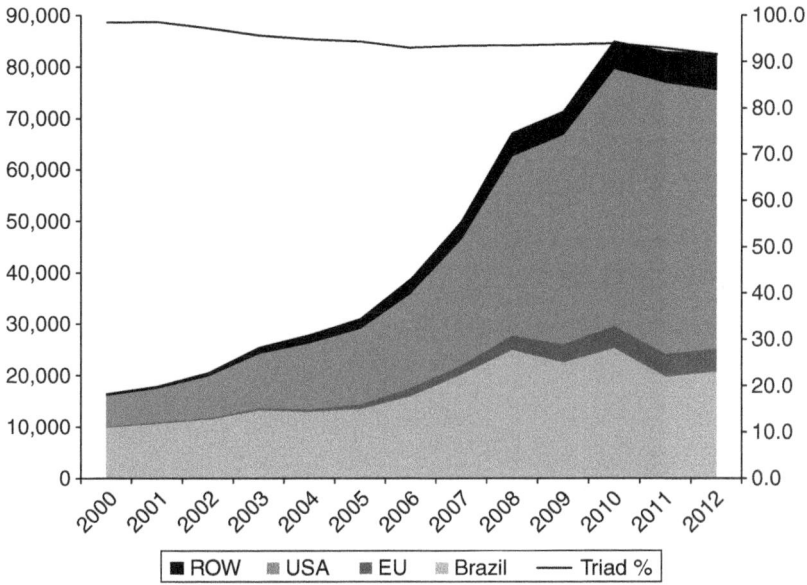

Figure 1.1 Fuel ethanol production, million litres, 2000–2012, and triad share, per cent of world total

Source: International Sugar Organisation Ethanol Year Books, various years.

about 90 per cent of the world total. With biodiesel the dominance is not so great, but is still over 70 per cent.

Biofuels – the generation game

The second distinction to explain is between first generation and advanced biofuels. The key distinguishing feature here is between feedstocks that can be used as food for humans and feedstocks that do not have such end uses. First generation biofuels are sometimes referred to as conventional biofuels. These are derived from feedstocks such as sugarcane, sugarbeet, corn, wheat, soybeans, palm oil, rapeseed, castor oil – all of which also have uses, either directly as food for humans or indirectly as animal feed. All of these feedstocks also require land for cultivation – the significance of which will be explored more fully below and in Chapter 9.

Advanced biofuels are, in turn, split further into higher generations. Second generation biofuels involve a range of inputs which do not compete directly with food uses. These include non-edible parts of food

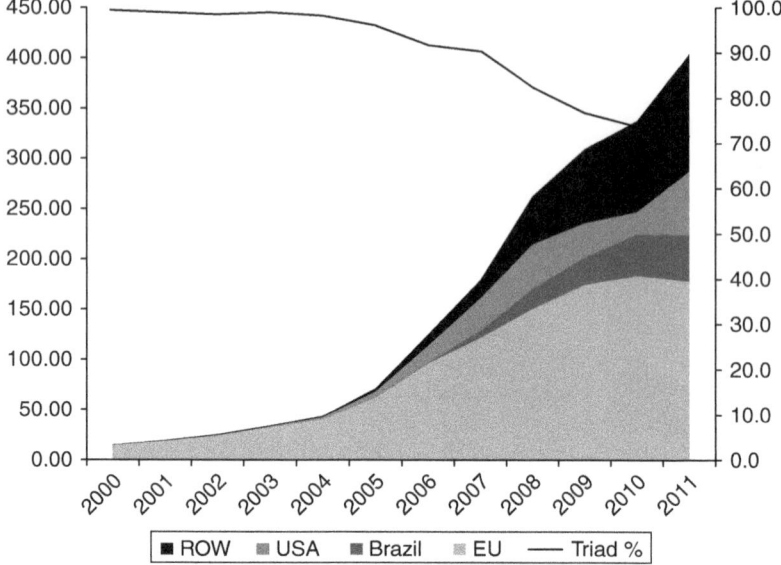

Figure 1.2   Biodiesel production, thousand barrels per day, 2000–2011, and triad share, per cent of world total
Source: US Energy Information Agency.

crops, for example, grain stover (stalks, leaves, husks, and so on), animal fats and the similarly waste elements of forestry. This category also includes non-food crops, such as certain types of grasses, which can be grown specifically for biofuels. That said, because this type of second generation feedstock requires land for its cultivation, some of the potential problems with first generation biofuels could still occur with second generation biofuels derived from such feedstocks. Another source of biodiesel is recycled cooking fats and oils. In this book we consider this to be a second generation biofuel because, as it is derived from waste products being recycled, its use for biodiesel no longer competes with its use as food.

There are, typically, two distinct definitions offered for third generation biofuels. One, specifically, describes biodiesel derived from algae. There is, however, a broader definition of third generation biofuels, an excellent summary of which is provided by Biopact[8] (see also Liew *et al.*, 2014). Second generation biofuel feedstocks involve bioconversion – the derivation of biofuels from the processing of a range of feedstocks. The Biopact definition identifies third generation biofuels as those derived from feedstocks which have been subject to 'advancements made at source'.[9] That is to say, there has been some adaptation made to the feedstock grown, prior to being harvested and converted into biofuel. Specifically, third generation biofuels are derived from feedstocks which have been *designed* as energy crops, with higher yields and improved bioconversion.

This latter point is very important, because it can lead to reduced production costs for biofuels, improved biofuel yields from feedstocks and so on. This development is, in part, a response to the fact that cellulosic biomass is a common feedstock type for second generation processes, but this has relatively high conversion costs because it is harder to break down than the sugars, oils and even the starch in first generation feedstocks (FAO, 2008: 18). The Biopact website cites evidence where plant-breeding efforts are leading to the development of feedstocks which already contain the enzymes required to break them down to produce fuels, making the process even easier and more cost-efficient.

Some studies also identify a fourth generation of biofuels. These are based on Utopian feedstocks which are capable of delivering a carbon-negative outcome (even the best renewable energy sources can only ever be, at most, carbon-neutral). There are two distinct stages of technological challenge with these biofuels. The first, on which scientists are beginning to deliver results, is to develop biomass crops capable of storing much more carbon than standard varieties. The Biopact website reports this is already being achieved with, for example, varieties of eucalyptus. The greater technological challenge comes at the stage of the conversion of these feedstocks into biofuels, where the carbon released is then captured and stored. Carbon capture and storage (CCS) technology is the key to delivering the carbon-negative outcome, and also offers benefits to the burning of fossil fuels, but successful commercial development remains elusive (see also Milne and Field, 2012).

In advance of detailed discussion in Chapter 4, we note here that US policy introduces a note of confusion into this standard classification of biofuels. It defines advanced biofuels in terms of the greenhouse gases (GHGs) emission reductions a particular biofuel delivers. Thus Brazilian sugarcane-based ethanol, as defined in the 2007 Energy Independence and Security Act (EISA), is considered to be an advanced biofuel, based on its emissions reduction performance; notwithstanding the fact that, based on an agricultural feedstock, it conforms to the general understanding of a first generation biofuel.

A feature of first generation biofuels is that the production processes are well-known and long-established commercially. Given the multiplicity of feedstocks and technology pathways which can deliver biofuels, however, there is a commensurately large degree of variation in production efficiency, costs, energy outputs and emissions from first generation biofuels. The technologies required to bring advanced biofuels to market are, meanwhile, at various stages of development. In the US, in particular, the EISA, analysed in detail in Chapter 4, has helped bring small quantities of cellulosic ethanol to market. Meanwhile, more or less all of the biofuels produced in the UK, for example, are derived from waste products. These successes, however, remain on a relatively small scale at the time of writing, compared with the total volume of first generation biofuels delivered to market.

A key issue which arises from this dominance of first generation biofuels – and a theme running throughout this book – is the fact that first generation biofuels have the potential to produce a range of downsides, which policy-makers then have to try to manage. It is possible that the production of feedstocks for biofuels could affect the price of food products, affect the price of animal feeds, cause significant ecological and ecosystem damage, produce greater emissions of GHGs than the fossil fuels they are replacing, and trigger changes in the use of land around the world which could add to all of these problems. This, in a nutshell, would be biofuels done (very) badly.

Herein lies one of the great challenges for policy-makers – not all biofuels, not even all first generation biofuels, trigger these side effects; therefore a policy should, ideally, be sufficiently nuanced and targeted that it can promote biofuels done well, whilst excluding or discouraging biofuels done badly. We introduce the policies currently in place in Part I of this book. A key motivation for developing advanced biofuels is to bring to market biofuels which avoid, or at least have lesser downsides than first generation biofuels. As a result, both EU and US policies are seeking, simultaneously, to promote advanced biofuels whilst trying to contain the downsides of first generation biofuels. This bifurcation of EU and US policies is analysed in detail in Chapters 3 and 4. In Chapter 7 we analyse how domestic policies have not yet proved vulnerable to external pressures arising from the possible impact of biofuels on land use and on food prices. Then, in Chapter 9, we analyse these two policy challenges in detail.

Why biofuels, why now?

Transport biofuels have been around for a very long time. Before the end of the 19th century, Rudolf Diesel demonstrated his new engine using a fuel based on peanut oil, whilst a few years later Henry Ford designed the Model T to run on ethanol. Subsequently, however, the emergence of cheap oil as a transport fuel displaced biofuels. Only in Brazil were extensive and continuous efforts made to develop biofuels – principally ethanol derived from the vast sugarcane resources available (see Chapter 2). The first oil crisis of 1973/74 saw countries begin to look once again at biofuels, but it was the turn of the 21st century that saw policy-makers promote the rapid and substantial expansion of biofuels production and use. As of early 2014, 34 countries plus the EU have either current or planned biofuel mandates, with over half having multiple mandates, mainly consisting of separate targets for ethanol and biodiesel.[10]

Throughout our fieldwork interviews and meetings, three key factors were identified as being behind this recent policy drive for biofuels: energy security, rural development and climate change mitigation. Their relative importance in each country has varied over time, as explored in Chapters 2–5. Moreover one interviewee, a senior civil servant in the European Commission sceptical of biofuels' ability to deliver what policy-makers desire of them, argued that the fact that there are three policy drivers is important politically. If one driver is criticised, supporters of biofuels can always point to one or both of the other factors as supporting the policy drive. At this stage, we introduce each briefly.

Energy security

Global road transport systems are almost entirely dependent on oil-based fuels. There are, however, only a relatively small number of countries with significant supplies of oil. This creates an economic dependency of the many importers upon the few exporters. Into this simple equation, however, comes politics. One of the simplest but most eloquent statements on this came from an interviewee in one of the US government agencies, who observed that 'we buy an awful lot of oil from people who don't like us very much'. Indeed, the biggest user of fossil fuel in the world is the US Department of Defense, The Pentagon. Energy security is not purely about oil-centric geopolitics, however. It also reflects the dependence of nations' energy matrices in general, and transport system in particular, on fossil fuels, which are finite in supply. As a result, notions of energy security can embrace diversification of fuel sources *and* fuel types.

This distinction is seen clearly in Chapters 2–4 where, at different times, policy-makers in our focus cases have drawn upon different aspects of the energy security argument. Noteworthy in this is the development of EU policy, whereby policy embedded a de facto import requirement for biofuels and biofuel feedstocks. When asked directly about the EU's interpretation of energy security, given this import requirement, several interviewees made the point that, by diversifying types of transport fuel, diversification of source countries is also achieved: this is how the EU is interpreting energy security. Various EU policy documents have referred to the geopolitical dimension of oil, but it has had nothing like the prominence and priority as a goal of biofuels policy, as it has in both Brazil and the US at different times. In both of these cases, promoting energy security has been seen in terms of increasing domestic energy production, via domestically produced biofuels, which can substitute for imported petrol and diesel.

Some interviewees in the EU, however, saw EU policy as running counter to energy security concerns. This arises from the balance of petrol and diesel in the EU transport fuel mix and the relative domestic production levels of the two biofuels and their respective feedstocks. Specifically, because the EU is already an exporter of petrol, production of ethanol is more likely to enhance petrol exports than energy security (when defined in terms of substituting for imported oil-based fuels). On the other hand, the EU is a net importer of diesel, as a result of which imports of biodiesel or feedstocks with which to produce biodiesel merely substitute for other types of import – they do not reduce significantly the EU's dependency on fuel imports.[11] Given this paradox, the importance of being able to define energy security in terms of diversification of fuel and country sources becomes clear.

There is, however, a further dimension to consider. As officials in the European Commission pointed out to us, from a technical point of view the EU could produce most or all of the mandated biofuel volumes laid out in the policy – but to do so would be so disruptive to food markets as to be foolish even to contemplate. This implies, therefore, that whilst defining energy security to embrace the diversification of imported energy types and source countries may appear a convenient way around the petrol/diesel export/import issue, there are wider concerns that point to the merits of a policy which does indeed take a broader perspective on energy security.

A further dimension to the question of defining energy security as diversifying source countries is the renewable nature of biofuels. Specifically, feedstocks suitable for conversion into biofuels can be grown in countries across the world. As a result, feedstocks and biofuels can be sourced from more countries, including those seen as more stable and reliable trade partners. For transport fuel specifically, the renewability of feedstocks also helps address the problem of the depletion of oil reserves. As one US interviewee put it, 'when did Noah build his ark? Before it started raining'.

Rural development

Rural development is a term which, like energy security, is given different meanings in our three cases. The first challenge is to define 'rural'. A widely used definition is that of the Organisation for Economic Cooperation and Development (OECD), which has a two-stage construction. First, at the NUTS 5 level (local municipalities and communes),[12] a rural *community* is defined in terms of population density below 150 inhabitants per square kilometre. Based on this, there are then three types of *region*, defined usually at the NUTS 2 or 3 level. A *region* is predominantly rural if over 50 per cent of the population lives in rural *communities*. A region is significantly rural if 15–50 per cent of the population lives in rural areas. A region is predominantly urban if less than 15 per cent of the population lives in rural communities. Significantly, this definition refers to population density rather than economic activity: that is, 'rural' is not congruent with 'agriculture'. That said, policies referred to as 'rural development', such as those of the EU, tend to be focused primarily on participants in farming. Moreover, the production of feedstocks for first generation biofuels, and several feedstocks for second generation biofuels, is land based. This offers alternative income opportunities for farmers, foresters, etc. With a lot of biofuels production located close to the production of the feedstock, this brings additional industrial jobs into rural areas.

One perception of biofuels policy, however, is that it is an agricultural support policy by another name. As such it would be aligned not so much with what the EU and others now promote specifically as rural development policies but, rather, the traditional agricultural support policies seen across the developed world, including the EUs Common Agricultural Policy (CAP). In the EU, interviewees across different Directorates-General of the European Commission gave diametrically opposing views about whether biofuels policy was, in part or in full, a new form of agricultural support policy. In particular, was the biofuels push related to reform of the EU sugar regime, to offer an alternative market for sugarbeet growers? When, previously, we researched that reform (Ackrill and Kay, 2009: 2011a), no interviewee identified a causal link between sugar reform and biofuels. Only when we later undertook research on the EUs biofuels policy did some interviewees then refer back to sugar.

Ultimately, this issue remains unresolved in the EU context. In the US, however, the views expressed were clear. As several interviewees put it (civil servants within government agencies, committee staffers within Congress and industry insiders), this is a policy for 'Big Ag'. Empirically also, the links are much stronger and more evident between biofuels and 'traditional' agricultural policies. Specifically, US biofuels policy has been dominated thus far by the production of ethanol derived from corn (maize). At key points in the evolution of US corn policy, clear connections are evident between developments in agricultural policies and markets, and ethanol policy. These are explored in more detail in Chapter 4. In Brazil, too, as analysed in Chapter 2, a key ethanol policy introduced in 1975 was influenced in part by the lobbying of sugarcane growers. Energy security might have been the primary concern, but the policy's ability to offer support to farmers was also a factor in its development.

Climate change and emissions reductions

However energy security is defined, biofuels do substitute for fossil fuels. Equally, however rural development is defined, the production of biofuels feedstocks and the often rural location of biofuels production facilities do provide farmers with an alternative outlet for crops (whether food crops or not), and do provide non-farm employment opportunities in the biofuels segment of the supply chain. Do biofuels, however, deliver GHG emissions reductions? The answer is that not all do, and even those that do reduce emissions do so to varying degrees. It is thus with the climate change mitigation driver of biofuels policies that the greatest ambiguities lie. This, in turn, provides the space for conflict and disagreement over the merits of biofuels.

The burning of fossil fuels releases GHGs into the atmosphere. By replacing fossil fuels with biofuels, the intention is to reduce the emissions from transport. This is not the largest source of emissions currently, but transport is forecast to be the major source of emissions growth in the coming decades. The challenge with biofuels in this regard, however, relates primarily to their production. In short, the production of biofuels – especially but not only first generation biofuels – involves activities which also generate GHG emissions. As discussed below, this can have a profound impact on the net emissions performance of different biofuels. Moreover, the calculation of these emissions faces major challenges. For example, every combination of feedstock and technology pathway will generate different emissions. Moreover, there is a hugely controversial debate over potential knock-on effects beyond the biofuels supply chain. On this point, we introduce below the concept of Indirect Land-Use Change (ILUC), and analyse it further in Chapter 9.

As explored in the following chapters, the emissions and climate change driver of biofuels policy has been used differently in the policy narratives of our three focus countries. In Brazil, it has risen in importance in recent years, as the country develops economically whilst also wishing to develop a global leadership role in environmental and climate change-related matters. In the EU, biofuels policy has a strong link back to the EU's emergence as a global leader in climate change policies in the 1990s. In the US, however, climate change and other environmental dimensions have played, at most, a minor role. Indeed, on our second visit to Washington DC, in 2011, we were told by a civil servant in one government department, only half-jokingly, that only in the last year had a climate change dimension to biofuels policies begun to be whispered. Given the importance of GHG emissions to the biofuels debate, we consider in more detail below the issues and challenges which underpin this aspect of biofuels policies.

Just these three drivers of biofuels policies?

Some authors identify a range of factors driving biofuels policy, beyond these three. In an extensive review of the literature, including governments' policy documents, IEA Bioenergy (2009: 64) additionally identifies, for example, the environment, agricultural development (as a distinct category from rural development) and technological progress as biofuels policy drivers. In our interviews and meetings across the EU, US and Brazil, the three drivers identified above dominated discussion. Moreover, the aforementioned 'additional' drivers were brought into the discussion only as elements within the three principal policy drivers.

Environmental considerations, typically, were discussed as part of climate change. This manifests itself, for example, in emissions and biodiversity aspects of sustainability criteria (see Chapters 3, 4 and 8). Rural development, as indicated above, is a difficult concept to pin down, as different countries and interest groups can have very different definitions and understandings of what it means, both in general and in the context of biofuels. As a result, the distinction made by IEA Bioenergy, 2009, between rural and agricultural development, can be accommodated by a broad understanding of the concept of rural development. As discussed in Chapter 2, Brazil has utilised biofuels the most in the broader sense of contributing to economic development (understandably so, given its emerging economy status, in contrast to the EU member states and the US). Interestingly, technological progress was referred to by several of our interviewees in the context of rural development. Specifically, biofuels offer 'industrial' jobs in rural areas but which, also, necessitate technological developments – utilised, significantly, in that rural context.

Related to the previous point about Brazil, in discussions following the presentation of some of our work at a seminar at the World Bank in Washington DC, it was brought to our attention that, in the past, biofuels were also seen as a potential instrument of economic development in developing countries. This can be seen, to an extent, in our discussion of Brazilian policy in Chapter 2. Beyond this, however, the role of biofuels in economic development is something that warrants specialist and detailed analysis in its own right – going, as it does, well beyond the scope of our project and this book (see, for example, Mitchell, 2011; Clancy, 2013).

Arguably the clearest expression about the drivers of biofuels policy came from one of our Brazilian interviewees. Whilst he recognised the arguments based around the various policy drivers outlined above, ultimately he was absolutely clear that it was about stimulating domestic production: 'nobody is doing these mandates to import biofuels'.

Because we can...

Given the policy drivers identified above, the explanation for the 'biofuels frenzy' since the turn of the century is, in essence, very simple. First generation biofuels – both ethanol and biodiesel – can be produced using known technologies at a commercial scale, using available inputs, and can be blended with fossil fuels to power existing vehicle engines. There are limits to the volumes of ethanol that can be blended with fossil fuels and burnt in standard, unaltered engines, but the basic principle holds true. As a result, politicians and policy-makers are able not only to express concern over climate change, rural development and energy security, but can be seen to be doing something about these concerns.[13] It should also be added that, as indicated by several interviewees in all three cases, biofuels are a way of seeking to reduce transport-related emissions without having to alter, fundamentally, people's behaviour – they can still drive their cars as before. The alternative – of getting people to change their mobility habits – would be fraught with political difficulty. In this regard, biofuels were particularly attractive.

A central theme of Part II of this book is what happens when the downsides to this are revealed. As one interviewee, a senior official in a US government department, said, the best biofuel is always the next one, because we do not yet know its downsides. In terms of the three drivers, this applies most powerfully to the question of emissions. Moreover, because first generation biofuels are derived from feedstocks that can also be used to produce food, their use for biofuels has potentially significant implications for food production and food prices which, as analysed in Chapter 9, is not always immediately obvious ex ante.

As a result, EU and US policies are taking a two-pronged approach which, in both cases, is leading to increasingly bifurcated policies. In both countries there are considerable efforts to promote the production of advanced biofuels. This helps avoid food v. fuel conflicts, and can limit land-use concerns. This, in turn, helps to contribute to what is, on average, a far better GHG emissions-reduction performance than first generation biofuels. In the meantime, both countries have in place 'sustainability criteria'. These seek to ensure that the first generation biofuels entering the transport fuel mix are derived from feedstocks which minimise, as far as possible, the negative emissions and environmental consequences arising from their production and use. To add to the policy challenges, these sustainability criteria, whilst intended to minimise one set of downsides from biofuels, can potentially cause another if they represent barriers to trade under WTO law (see Chapter 8).

Chapters 2–4 put these different ideas into their national contexts. Part II of the book analyses the various policy challenges in more detail, challenges both to policy-makers and policy-making processes. This refers to the implications of using different feedstocks, different technology pathways, having different climatic and agronomic conditions in different countries, the different 'generations' of biofuels and the implications for policy-makers regarding the scale and rate of expansion of biofuels markets. We may be doing biofuels, especially first generation, because we can; we seek in this book to determine, analytically, parameters which can inform decisions about whether, and in which situations, we should be doing biofuels.

Challenges facing policy-makers – An introduction

Underpinning the arguments in this book is the notion that there are no such things as good biofuels or bad biofuels; only biofuels done well and biofuels done badly. In this section we introduce some of the key challenges that policy-makers face, as they seek to design or reform policies to encourage the expansion of biofuels markets, seeking to ensure they promote biofuels done well and avoid biofuels done badly (we analyse these issues in detail in Chapter 9). To put some of these issues into context, we begin by considering briefly the calculation which determines the GHG emissions performance of different biofuels. For this, we consider the calculation specified by the EU in Annex V, Section C, of the 2009 Renewable Energy Directive, or RED (EU, 2009a). See Annex V of the RED for further details of the GHG emissions calculations; and Chapter 3 of this book for analysis of EU policy. By so doing, we seek to illustrate the complexity of the calculations required as part of the implementation of biofuels policies. This also underpins our analysis of ILUC in Chapter 9.

Calculating GHG emissions from biofuels

Greenhouse gas emissions from the production and use of transport fuels, biofuels and bioliquids are, for EU policy purposes, calculated as:

$$E = e_{ec} + e_l + e_p + e_{td} + e_u - e_{sca} - e_{ccs} - e_{ccr} - e_{ee}$$

Where:

E = total emissions from the use of the fuel;
e_{ec} = emissions from the extraction or cultivation of raw materials;
e_l = annualised emissions from carbon stock changes caused by land-use change;
e_p = emissions from processing;
e_{td} = emissions from transport and distribution;
e_u = emissions from the fuel in use;
e_{sca} = emission saving from soil carbon accumulation via improved agricultural management;
e_{ccs} = emission saving from carbon capture and geological storage;
e_{ccr} = emission saving from carbon capture and replacement; and
e_{ee} = emission saving from excess electricity from cogeneration.

Emissions from the manufacture of machinery and equipment shall not be taken into account.

This figure, E, is subject in the EU to an initial GHG emissions-reduction condition of 35 per cent lower emissions than fossil fuels, a figure which rises subsequently to 50 per cent and 60 per cent for different processors (see Chapter 3 for details):

$$(Ef - E)/Ef \geq 0.35$$

Where:

Ef = emissions from the fuel comparator

Emissions are expressed as gCO_{2eq}/MJ – grams of carbon dioxide equivalent, per megajoule of fuel. In addition to carbon dioxide, the calculations should also include nitrous oxide (to be included at a factor of 296:1 compared with carbon dioxide in emissions calculations), and methane (at a factor of 23:1). Annex V of the RED then provides further details on individual variables in the above equation. In addition, it provides data for both 'typical' and 'default' greenhouse emissions for each variable in the equation used to calculate E. Where producers believe their biofuel offers a more favourable value than one of the default values, they can submit evidence to that effect.

One particular note of explanation, important in the analysis through the rest of the book, involves an equation to estimate 'annualised emissions from carbon stock changes caused by land-use change':

$$e_l = (CS_R - CS_A) \times 3,664 \times 1/20 \times 1/P - e_B$$

Where:

e_l = annualised greenhouse gas emissions from carbon stock change due to land-use change (measured as mass of CO_2-equivalent per unit biofuel energy);

CS_R = the carbon stock per unit area associated with the reference land use (a measure to include both soil and vegetation). The reference land use shall be the land use in January 2008 or 20 years before the raw material was obtained, whichever was the later;

CS_A = the carbon stock per unit area associated with the actual land use (also a measure to include both soil and vegetation). In cases where the carbon stock accumulates over more than one year, the value attributed to CS_A shall be the estimated stock per unit area after 20 years or when the crop reaches maturity, whichever the earlier;

P = the productivity of the crop (measured as biofuel energy per unit area per year);

e_B = bonus of 29 gCO_{2eq}/MJ biofuel if biomass is obtained from restored degraded land (after which follows an extensive discussion of what qualifies for this bonus).

Only one interviewee, an official within the European Commission, volunteered any opinion on this equation. They saw the 29 gram bonus as lacking any scientific basis, disconnected from the costs of restoring degraded land. Instead, they expressed the opinion that it was included as a political compromise, introduced as part of a deal over a separate problem we discuss next – ILUC.

Land-use change

Land-use change (LUC) is one of the most controversial aspects of biofuels. LUC comes in two variations, direct (DLUC) and indirect (ILUC), with the latter being particularly controversial. DLUC can be considered as occurring where land previously used for another purpose is now used to grow feedstocks for biofuels production. It is critical to understanding the recent pressures on EU policy, as explored in Chapter 3, to recognise that the earlier equation for land-use change deals specifically with DLUC. ILUC, analysed in Chapter 9, is very different and is thus being dealt with very differently by policy-makers.

To illustrate how LUC works, a few simple examples will help. By way of introduction, in the EU the European Commission defines LUC in terms of changes between any two of the six land uses classified by the Intergovernmental Panel on Climate Change (IPCC) (forest-land, grass-land, cropland, wetlands, settlements and other land), supplemented by a seventh use, perennial crops (which straddle cropland and forest land).

If, for example, a German farmer who previously grew oilseed rape for food uses continues to grow oilseed rape, but now sells it to a biodiesel refinery, there is no land-use change at all. Let us call this crop-use change, on a given area of cropland. Alternatively, if this farmer instead brought land previously under any of the other land-use categories above into 'cropland', specifically to produce the feedstock for biofuel, that would be DLUC. This would also apply if the new crop is a non-food crop grown specifically for second generation biofuels.

What about a case in between the previous two, where our farmer used to grow wheat for food but now, on that piece of land, changes both the crop grown (say, to rapeseed) and the end use (for biofuel)? This would be an intra-cropland change in land use. But what about the change in the end use of the crop now being grown? Even a cursory glance at research papers confirms there is no consensus as to whether or not this should be referred to as an example of DLUC.[14] The EU emissions calculation shown above does not consider it as part of any DLUC effect. That said, if we do consider it as an example of DLUC, this helps create a conceptual bridge to the rather more thorny issue of ILUC, which we now turn to.

To consider what ILUC is and what it involves, let us return to our hypothetical German farmer. If this farmer and enough colleagues decide to switch out of wheat as a food crop, into oilseed rape for biodiesel production, the price of wheat might go up on domestic markets, even on international markets. As a result, farmers in, say, the Ukraine might plough up pasture land in order to grow more wheat. This would be an example of ILUC, whereby EU biofuels policy has a *direct* effect on land use within the EU, but which has in turn had an *indirect* effect on land use elsewhere. This latter effect has occurred via the price linkages between commodity markets and between countries. The emissions consequences of this, some argue, should be included in the calculation of E for the biofuel produced using our German farmer's (and his colleagues') rapeseed. The challenge of ILUC is, first, knowing that these land-use changes are occurring (this could be happening many thousands of miles away); and, second, that they are happening as a direct causal consequence of our first farmer's decision back in Germany to switch crops, and then to sell the new crop for biofuels.

Thus if the Ukrainian farmers are producing more wheat *solely* because of the higher international wheat price, and so long as that higher wheat price has been caused *solely* by the German farmers switching out of wheat and into oilseed rape in order to supply a bio-refinery and so long as the land-use change required to produce more wheat in the Ukraine *triggers* the release of carbon stored in the soil, *then* that released carbon should be included in the calculation of the GHG emissions attributable to the German rapeseed-biodiesel. Therein lie some of the multiple challenges to those seeking to estimate ILUC. In our hypothetical and, in truth, highly simplified example:

- It is assumed that the German farmers are producing more oilseed rape as a result of increased demand for biodiesel in the EU, because of the biofuel policy. They may be producing more oilseed rape for other reasons (instead of or as well as biofuel-related), for example, increased demand for rape oil as a food product; or in response to higher international rape prices following a poor harvest in another part of the world, perhaps as a result of an extreme weather event.
- It is assumed the rise in wheat price arises entirely because of lower production resulting from the German farmers switching to oilseed rape to supply the bio-refinery. This, in turn, assumes that, first, the first point above applies; and, second, that as a result, enough farmers switch out of wheat in Germany to affect the world price of wheat by

enough, to influence the production decisions of Ukrainian farmers. It thus also assumes that there are no other factors, anywhere in the world, which have contributed to the higher world wheat price.

- It is assumed that the only reason for the revised production decision of Ukrainian farmers is the higher wheat price, itself conditional on both of the above points holding. What if, for example, a drought in Russia or flooding in Canada has caused, or at the very least contributed to, the rise in wheat prices? What if wheat prices are rising as a result of rising global demand, driven by one or both of rising global population and global income? Then, at most, only a portion of the estimated ILUC-related emissions should be attributed to biofuels via the calculation of E... which begets the question, how big should that portion be?

Thus, even if ILUC occurs, estimating its magnitude is an enormous challenge. The fundamental reason for this, implicit in the preceding hypothetical examples, is that it cannot be observed directly. It requires estimation via economic models, because it occurs as a result of economic factors, essentially price transmission effects, which connect various biofuel, food and land markets. ILUC occurs when, as a result of farmers in one location deciding to produce feedstocks for biofuels rather than for food, other farmers elsewhere bring land into the production of food as a result of price changes caused by the production and market decisions of the first farmer, who has taken their output away from a particular food market and sold it instead in a biofuel market.

DLUC occurs all the time, as a result of the complexities of the economic inter-connectedness of markets. Whilst there are established values and methodologies for calculating DLUC, as shown above, the same cannot be said of ILUC. If carbon emissions from ILUC are to be attributed to the production of the feedstocks going into biofuels, assumptions have to be made about key economic relationships, as outlined above. As a result, however, every assumption represents a potential flashpoint for debate and disagreement between those in favour of and those opposed to biofuels.

Of our three focus countries, only in the EU has this created sufficient controversy to influence policy-makers – as discussed in Chapter 3. In the US, with biofuels dominated by domestically produced corn ethanol, the principal land-use effect is DLUC. The consensus across the interviews we conducted with officials in government agencies and staff within Congress, in addition to representatives of the biofuels sector, was that even if ILUC was occurring as a result of US policy, its scale was minimal. In Brazil, the situation is more complex. Whilst government officials felt that ILUC in Brazil was minimal, there are others who feel it is significant and serious – as evidenced, for example, by continuing deforestation in the Amazon region, and by land-use changes in other biomes, notably the Cerrado.

Interviewees in Brazil identified another factor affecting markets and production which might trigger land-use changes – the considerable differentials in land prices in different parts of the country. With land on the frontier of agriculture typically one-eighth to one-tenth, but as little as one-thirtieth, of the land currently being farmed, the incentives to buy land on the frontier are significant. Landowners may buy land and then, indeed, subsequently use the land for a few years to grow crops for biofuels production, but the decision to buy the land may well be motivated by price-related factors, such as land speculation, with longer-term plans for the land perhaps totally unrelated to biofuels. Again, we see the challenges facing those who seek to model accurately the ILUC effects of biofuels and biofuels policies are many and complex.

Biofuels, energy balance and competitiveness

The foregoing discussion has shown how difficult it is to estimate with a degree of accuracy the GHG emissions performance of different biofuels – and thus the potential savings relative to fossil fuels. Indeed, some argue that by the time ILUC has been factored in, biofuels can generate greater emissions than fossil fuels (see Chapter 9). This challenge is compounded by the fact that every feedstock, every technology pathway to conversion into biofuel and thus every combination of these two will deliver different GHG emissions reductions.

Moreover, GHG emissions are not the only indicator or benchmark for judging the performance of biofuels and their likelihood of adoption (in addition to their ability to deliver greater energy security or contribute to rural or agricultural development). Consideration can be given to their different energy balances (comparing the energy used in their production with the energy generated by their use), again not only with each other but also against fossil fuels. As for adoption, there is the question of the relative price, allowing for the fact that (as noted above) the energy content per unit volume of different biofuels varies compared with fossil fuels.

Taking first the question of GHG emissions changes relative to fossil fuels, the challenge is to summarise the almost infinite number of estimates available, based on feedstock, biofuel-type, technology pathway, etc. – and given the fact also that technologies and production

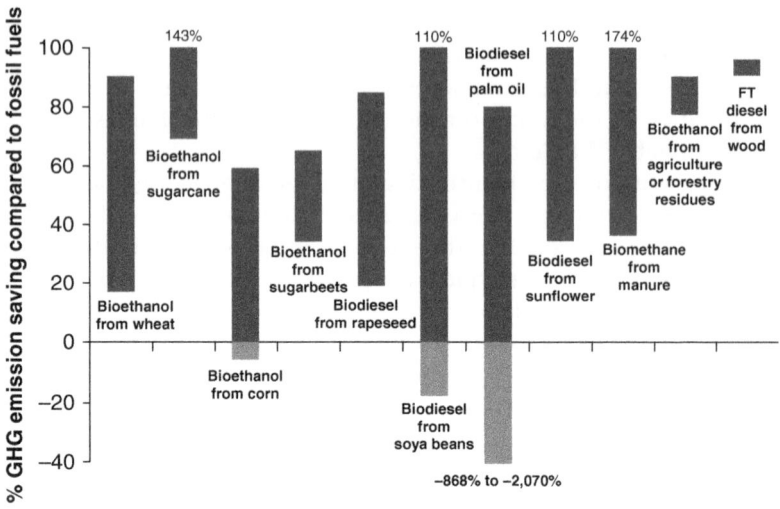

Figure 1.3   Greenhouse gas savings of biofuels compared to fossil fuels
Source: UNEP (2009: 54).

processes are continually evolving, with the resulting biofuels superseding the emissions performances of biofuels even produced in the same country and the same factory. A really useful summary is provided by UNEP (2009: 54) (Figure 1.3).

This shows the range of potential estimated emissions savings, relative to fossil fuel, for a selection of biofuels and feedstocks. The highest maximum emissions saving, from sugarcane ethanol, is higher than the estimates of the second generation biofuels presented. Moreover, the minimum estimated emissions performance for sugarcane ethanol exceeds the maximum possible performance for the largest single source of ethanol – corn. Studies also suggest that the maximum performance figure for corn ethanol is only achievable with adoption of the very latest technologies (see UNEP, 2009: 53). As we shall see in Chapter 4, US policy 'grandfathers' older ethanol production facilities, as a result of which the highest potential corn ethanol emissions savings overstate what US policy, in aggregate, is delivering.

The UNEP data also show that whilst biodiesel derived from palm oil can deliver up to an 80 per cent emissions reduction relative to fossil fuels, it is also capable of delivering emissions of 900 per cent, or even over 2000 per cent, greater than fossil fuels. These two figures are estimated for land conversion from, respectively, rainforest and peat forest. As discussed in Part II, both of these scenarios are plausible, but they are not a necessary or inevitable feature of producing biodiesel from palm oil. As we shall see, in Chapter 8 in particular, the introduction of sustainability criteria is, in part, motivated by a desire to avoid such damaging feedstocks being used in biofuel production.

On the other side, one reason why Brazilian sugarcane ethanol is so effective at delivering emissions reductions is that, in a lot of factories, the energy required comes from burning the bagasse – the pulp resulting from the crushed cane. In some cases, technology even captures and uses water and steam from the conversion process, to help produce electricity at two or three stages in the factories' ethanol production processes. As a result, the fossil-fuel energy balance for sugarcane ethanol (the energy contained in the biofuel, per unit of fossil fuel used in the production and processing of the biofuel) is particularly high. It also explains why UNEP estimates a maximum emissions reduction performance of over 100 per cent, with surplus electricity being sold into the national grid.

Given the range of factors and estimates for biofuels' energy balance, we can give only some general indicative figures, in this case taken from Brown and Brown (2012: 163). Sugarcane ethanol is shown as having an energy balance of 8 (some of our interviewees suggested it could be as high as 9 or 10). That is to say, sugarcane ethanol delivers 8, 9 or even 10 times the energy used in its production. Palm oil biodiesel has a figure of 9, whilst cellulosic ethanol can range from 2 to 36 (which serves to illustrate how important advanced biofuels could be, potentially, in the future). Biodiesel derived from waste vegetable oil has a value of 5–6, whilst rapeseed biodiesel (widely produced in the EU) comes in at about 2.5. Corn ethanol, dominant in the US, is estimated at 1.5. In terms of the baseline, it is crucial to understand that the production of fossil fuels itself uses energy. Thus, the energy balances for petrol and diesel are both below 1.

The final issue we consider in this section is the competitiveness of biofuels compared to fossil fuels. Once again, the theme running through this chapter, of the considerable variation across fuel types, feedstocks and technology pathways makes it impossible to deliver a simple, clear, estimate of this. Analysis is complicated further by the interplay of economic factors and markets, in some ways similar to the complications with estimating ILUC, outlined earlier. The key price comparison is between that of the fossil fuel and the biofuel. Fossil fuel prices can be volatile, susceptible to geopolitical instabilities as well as economic factors. Biofuels, meanwhile, compete for inputs with products destined for food markets on the supply side, whilst changes in oil prices will affect the demand for biofuels.

Indicative figures from the FAO (2008: 36) suggest that, for US corn ethanol, if crude oil was $60 a barrel, ethanol would be viable at corn prices up to $79.50/tonne. At $100 a barrel, corn prices could rise to $163/tonne and ethanol still be competitive. A widely quoted figure for Brazilian sugarcane ethanol is that it breaks even with oil priced at about $35 a barrel. Moreover, with the energy differential, ethanol has to be at or below 70 per cent of the petrol price at the pump for ethanol to be cheaper. Ultimately, as with much else in this chapter, there are such wide variations around these estimates, because there are so many different ways of producing ethanol and biodiesel.

Food v. fuel and food prices

Changes in the production of different agricultural commodities for the purpose of selling them for first generation biofuels production affect the supply of those commodities entering food markets and thus the price of those commodities. This, to the extent that it happens in practice, is the key trigger for ILUC, but it also has direct consequences for consumers. Once again, however, determining the magnitude of the contribution of biofuels feedstock to a given price movement is challenging. Moreover, a distinction needs to be drawn between commodity prices and food prices. This introduces a further dimension into the modelling and estimation of the impacts of biofuels – that of the transmission of commodity price shocks through the supply chain to (consumer) food prices.

In addition to food prices, the use of agricultural feedstocks for biofuels can also potentially affect the prices of animal feeds. This link is not straightforward, however. This is because several types of feedstock used to produce biofuels deliver by-products which can be used as animal feed. With some biodiesel feedstocks, the residue following the extraction of oils can be used to make a 'cake' for animal feed. Similarly, the residue left after corn and other cereals have been used for ethanol production can be used as an animal feed known as distillers grains with solubles (DGS). This can, further, be 'dried' (to give DDGS). As DGS, this can be fed wet to cattle close to the bio-refinery, avoiding the need to undergo the drying process and, thus, saving energy and costs (see Energy Future Coalition, 2007: 7).

In the US in particular, there have been significant debates between the ethanol industry and the cattle farmers over the impact of ethanol production on feed prices. On the one hand, the latter have argued that the diversion of corn into ethanol has had a significant impact on feed prices, whilst biofuel industry representatives have drawn attention to the price and market impacts arising from the increased production of cattle feed as an ethanol by-product. As well as these contrasting price effects, there are similar debates over the emissions consequences of animal-feed by-products resulting from biofuels production. Specifically, the issue has been raised as to whether the emissions calculation for a given volume of biofuel should be credited with emissions savings, achieved as a result of the double use of land – to produce feedstocks which deliver both biofuel and animal feed. We look at all of these issues, and more, in detail in Chapter 9.

Conclusions

Biofuels, as we have seen in this chapter, represent a complex set of products (even when we just analyse the two main types dominating markets currently – ethanol and biodiesel), and they raise a multitude of complex questions over what they deliver, beyond propulsion for motor vehicles. Collectively, biofuels are capable of delivering many benefits. If we are not careful, however, they are also capable of resulting in significant and potentially extremely damaging effects – economically, socially and environmentally.

The challenges for policy-makers are to try to ensure the biofuels delivered to markets have been produced in such a way as to maximise the benefits and minimise the downsides. This is an extremely difficult task, given the range of potential downsides that can result from biofuels. Yet the positives make the pursuit of 'biofuels done well' worthwhile. The policy challenges that policy-makers see biofuels as being a response to – energy security, climate change mitigation and rural development – are extremely complex. As a result there is no single solution, no 'silver bullet', that can address these challenges. Several of our interviewees declared that there was indeed 'no silver bullet', only a few years after biofuels were being presented as just that.

This change in the rhetoric, reflected also in the more formal policy discourse, is very important in the context of this book. Biofuels policies have been set up for a variety of reasons, the underlying motivations for which have not, in the main, altered significantly in the last decade. As a result, there has generally been a forward direction of travel in biofuels policies. It is to these policies that we now turn. In Chapters 2–4 we look at policies in Brazil, the EU and US, respectively. In Chapter 5, we review those policies via the three key drivers of policy identified here, looking for similarities and differences across our three cases. This sets the scene for Part II of the book, where we analyse key dimensions of policy – design, implementation, international aspects, and current and future challenges. Policy-makers have invested a great deal of political, as well as economic, capital in biofuels. The remainder of this book seeks to establish how successful this has been to date, and the challenges faced if this is not to be lost entirely.

2
Brazilian Biofuels Policy – An Introduction and Overview

Introduction

In this chapter we review the development of Brazilian biofuels policies. Through this chapter, a number of key features will be explored which will highlight just how different Brazilian policies are from EU and US policies. First, Brazil's ethanol (or 'alcohol') policies are fundamentally different from its biodiesel policies. Second, Brazil's ethanol policies have a much longer continuous history than those of either the US or EU. On the other hand, Brazil's biodiesel policy is newer than EU and US biofuels policies. A further difference is that, with ethanol in particular, there is more or less a free market in Brazil. We introduce these concepts here, returning to analyse them further in later chapters.

Ethanol in Brazil – The first 50 years

Ethanol is Brazil is derived almost entirely from sugarcane. Ethanol policy began, effectively, in 1933 with the creation of the IAA, the Institute of Sugar and Alcohol (Instituto do Açucar e do Alcool),[1] reflecting not only the importance of sugarcane for Brazil, but also the intimate interconnectedness of the markets for these two products. The context for the creation of the IAA was a period of disruption to sugar markets and, with it, economic difficulties in Brazil, principally the poor North and North East (NNE) regions of the country (Johnson, 1983). A combination of factors led to this disruption. Johnson (1983: 247) notes this sector had been in decline for some time, through a combination of rising production elsewhere and rising trade barriers against Brazilian exports. The Great Depression compounded the situation, triggering a highly damaging slump in prices (see, inter alia, Hutchinson, 1961: 208;

Johnson, 1983: 243; Schmitz *et al.*, 2002: 127). This was especially so in the NNE, given the already parlous state of the economy, with poor infrastructure and inefficient sugarcane production.

Thus Johnson (1983: 248) emphasises the role of the IAA in stabilising the domestic sugar price, whilst Puerto Rico *et al.* (2010: 1875) highlight the role played by ethanol in enabling producers to accommodate and offset the impact of fluctuating sugar prices. The IAA was given complete control of the market, regulating the amount of cane going for its various products (sugar, ethanol and molasses). The only exception was the cane used in the 'traditional sector' (Johnson, 1983: 247)[2], although the first mandate for blending ethanol into petrol, at 5 per cent, came slightly ahead of the IAA, in 1931 (Hira and Guilherme de Oliveira, 2009: 2451; Puerto Rico *et al.*, 2010: 1876–1877). The creation of the IAA reflected these economic and social factors, but also (and not unrelated) the establishment of a new dictatorship under Getúlio Vargas. He did away with elected regional and local representation, putting in its place appointed administrators (Hutchinson, 1961: 208). Thus the centralised control of markets given to the IAA reflected wider political values.

Hutchinson identifies 1888 as a key date in the history of cane production in Brazil and thus in the history of the nation, as it saw the emancipation of slaves. From that time through to the 1930s a shift occurred, from plantations (*engenhos*), run by *senhores de engenho* who both managed and owned everything and everyone, to the *usineiro*, a mill-owner or director who ran a central mill (*usina*), supplied by cane growers (*fornecedores*, the *engenhos* who went along with this new production structure). Over time, however, rather than remaining as millers to the surrounding cane growers, the *usineiros* expanded their landholding. This occurred partly in response to the larger scale of these new mills, the finer quality of production and the need for more production to ensure adequate return on capital.

This created 'a new type of plantation society, far larger and more complex than anything known before' (Hutchinson, 1961: 208). In short, the power of the *senhores de engenho* had been replaced with the power of the *usineiros*, over individual cane suppliers and, with it, political power at the local and state levels. Thus the IAA also enabled the balance of political authority in the NNE to be altered (see Hutchinson, 1961: 206–212). Data from Brazil's Ministry of Agriculture (MAPA)[3] indicate that, from the late 1940s to the mid-1980s, the cane supplied to mills was split roughly 50/50 between mill-owned production and that supplied by *fornecedores*.[4]

The economic problems faced in the NNE were compounded by a change in cane growing which was occurring by the 1930s, and which has continued to the present: a shift in the centre of gravity in cane production, from the NNE to the Centre-South (CS) regions (production in the CS overtook the NNE in 1951; Hira and Guilherme de Oliveira, 2009: 2451). 'Classic sugar cane plantations', *engenhos*, were established earlier on the coast of São Paulo state than the NNE, but only developed in inland São Paulo in the 19th century (Hutchinson, 1961: 203). São Paulo state is much better suited to cane growing, with better soil delivering higher yields, and with topography much better suited to mechanisation of planting and harvesting (aided still further by more recent, wider, economic changes, making for easier access to finance for mechanisation). The sugar industry in the CS was thus much more efficient that that in the NNE. In the early years of the IAA therefore, policies partly reflected the socio-economic differences (and, by implication, the political differences) between the two regions. Our interviews with senior government officials in Brasilia revealed that even today, the states of the NNE are over-represented in Congress relative to population, with a consequent impact on policy-making decisions, including those relevant to biofuels.

An important feature of IAA policy, again reflecting the wider political agenda, was one of import substitution. The disruption imported from turbulent world markets led to an inward-oriented IAA policy which, according to Johnson (1983: 253), was only just starting to be reversed by the late 1940s. Significantly, Brazil was one of the 23 signatories, or Contracting Parties, of the General Agreement on Tariffs and Trade in 1947. The willingness to produce for export increased notably from 1960, when the US closed its market to Cuban sugar following the Cuban Revolution, opening up significant new export opportunities (Hira and Guilherme de Oliveira, 2009: 2451), with IAA the monopoly exporter (Johnson, 1983: 247; see also Sheales *et al.*, 1999: 58). That said, as we shall see, a policy of promoting domestic self-sufficiency would return in the 1970s, especially for fuel (Puppim de Oliveira, 2002: 132).

The 1950s and 1960s saw cane production in the CS expand, with the easing of restrictions on installed capacity (Puerto Rico *et al.*, 2010: 1878). In 1959 *Copersucar*, the Cooperative of Sugar, Alcohol and Sugarcane Producers, was formed in São Paulo state, bringing together farmers, mills and refiners to support financing and marketing (Hira and Guilherme de Oliveira, 2009: 2451). During this period the industry was modernised and made more efficient, driven partly by greater surpluses and the need to improve external competitiveness. That said, this occurred mainly in the CS – the main players in the NNE were felt to be more interested in maintaining family power and maximising state support (Lehtonen, 2011: 2428).

Also during this period, a development was seen that would have a profound impact on Brazil just a few years later (Lehtonen, 2011). As the military government sought to modernise the economy, attracting multinationals to aid development of the motor industry played a central role. That said, some of the development money for the motor industry came as a loan from the International Monetary Fund (IMF), which contributed to the economic difficulties experienced subsequently (Rohter, 2010: 24–25). On the back of investment, however, from 1960 to 1976 the number of cars in the country rose by a factor of ten, oil consumption growth averaged over 16 per cent a year (80 per cent of this being imported) and economic growth averaged over 11 per cent (Lehtonen, 2011: 2427).

Proálcool – Brazilian ethanol production takes off

As a result of the factors outlined above, the oil crisis of 1973/74 hit Brazil hard, with the cost of oil imports rising from $606 million in 1973 to $2.6 billion in 1974 (Guan, 2010: 82). This threatened the 'Brazilian Miracle' of rapid economic growth. Brazil's oil import dependency also raised questions about the balance of payments and the country's ability to pay its rapidly rising import bill (Goldemberg, 2006: 1, puts the cost of oil imports at $4 billion, roughly equal in value to half of all Brazil's exports in 1973). Indeed, Brazil was already facing problems with its external debt, whilst in effect committing much of what it could import to capital goods and intermediate products, in order to continue industrialisation and deliver on the 1974–1978 National Development Plan (Teixeira de Sousa *et al.*, 2008: 121). Meanwhile, after a price spike, sugar prices fell back swiftly. Thus when producers of cane and of sugar lobbied for support, they were pushing at an open door. Discussions in mid-1975 considered financial assistance to the sugar industry (Hira and Guilherme de Oliveira, 2009: 2451) but, by the end of the year, a broader approach had been chosen.

Proálcool, Brazil's National Alcohol Programme, was launched in November 1975. It sought to reduce Brazil's dependence on imported oil and help bring balance to the balance of payments; diversify the use of sugarcane products and utilise the surplus available; expand the production of capital goods (cane processing being very capital-intensive) and reduce income inequalities at the regional and individual level (Puerto

Rico *et al.*, 2010: 1879). Proálcool would be managed by the Ministry of Industry and Trade, with the IAA still influential in price-setting (for molasses), promoting exports and overseeing issues related to technical specifications (Puerto Rico *et al.*, 2010: 1879).

Proálcool launched with two distinct variations, reflecting the distinctions between anhydrous ethanol (that which is blended with petrol) and hydrous ethanol (that which, with a small amount of water added, constitutes a fuel in its own right). All petrol sold was to be blended to include 10 per cent anhydrous ethanol, 'E10' (see also Carvalho, 1999). A second, voluntary, option was the use of hydrous ethanol (referred to as 'E100', the 5 per cent water notwithstanding) in engines adapted to take it (Goldemberg, 1996: 1). Yet such policy targets were not enough – also needed was the cooperation of the fuel companies and the car-makers. Teixeira de Sousa *et al.* (2008: 123) argue this was achieved through a combination of the power of the sugar-alcohol lobby and the pre-existence of 'technological and productive arrangements' (see below). Given the continuing high price of oil, the car-makers were probably not reluctant participants, notably in developing engines that could run on E100 and thus help maintain demand for new cars. Moreover (Hira and Guilherme de Oliveira, 2009: 2452), they hoped for state subsidies for this participation.

As well as the technologies and facilities already in existence for the processing of sugarcane into ethanol, there also existed a network of petrol pumps and petrol stations delivering 'super gasoline' which offered a ready infrastructure for conversion to ethanol (Goldemberg, 2006: 2).[5] At the other end of the supply chain, soft loans were made available for cane growers to establish distilleries, attached typically to sugar mills (Goldemberg, 2006: 2; Puerto Rico *et al.*, 2010: 1880). This helped address in a particularly direct way the spare capacity present at the time in the sugar industry, given the low price of sugar. As seen below, over time it would also help deliver economic efficiencies in the milling of cane and production of sugar and ethanol.

By 1978 petrol had evolved from an E10 blend to E23, but as well as delivering this higher blend percentage, there was no more spare capacity in sugar. Thus there was a debate about the future of Proálcool – was it intended as a short-term support measure for sugar, or a long-term means of promoting domestic energy supply and energy security? The future of Proálcool was, however, secured in 1979, as a result of the Iranian Revolution and another large rise in the price of oil (Hira and Guilherme de Oliveira, 2009: 2453).

Proálcool II – From boom to (nearly) bust to liberalisation and recovery

The year 1979 saw the start of a second phase of Proálcool. Since 1975, E100 production had been promoted on a voluntary basis, with incentives to buy cars capable of running on E100. What the car industry needed, however, was greater assurance that there would be a reliable supply of this (hydrous) ethanol. They also hoped fiscal incentives for the purchase of E100 cars would boost sales (Hira and Guilherme de Oliveira, 2009: 2453). New bodies were established to oversee ethanol production: CNAL, the National Council for Alcohol; and CENAL, the National Executive Commission (Rosillo-Calle and Cortez, 1998: 116). Key policy goals from 1979 were, thus, increasing the petrol blend to E25, ensuring ethanol prices were below petrol prices, guaranteeing minimum prices for producers, supporting an expansion of mill capacity and ensuring ethanol was available at petrol stations (Hira and Guilherme de Oliveira, 2009: 2453).

The response was dramatic. In January 1980, 1 per cent of new cars sold were E100 compatible. By December 1980, the figure was 73 per cent, added to by retrofitting standard engines (Hira and Guilherme de Oliveira, 2009: 2453). Moreover, all federal and state government cars were converted to alcohol (Sercovich, 1986: 152 and endnote 7). Predictably for a new technology, however, problems arose, such as high fuel consumption, engine ignition and corrosion (Sercovich, 1986: 152). As technical issues, however, most of these could be tackled. Separately, the rapid demand growth for E100 and E100 cars initially outstripped supply. Thus investment was made in more distilleries, many of which were stand-alone facilities, rather than being linked to cane mills. This, as we shall see, was subsequently to prove crucial for the industry.

Other issues had a longer-term impact. Rising sugar prices forced the government to increase the price of ethanol, to ensure sufficient cane went to ethanol. This, however, reduced the price advantage over petrol and hit demand for E100 cars. Even so, by 1985 about 96 per cent of new cars sold were ethanol-powered (Rosillo-Calle and Cortez, 1998: 116), with the E100 fleet making up about one-fifth of all cars in Brazil. Such an expansion, however, came on the back of substantial fiscal interventions. The beginning of the end of Proálcool, from the mid-1980s, came as a result of multiple, mutually-reinforcing, problems. The rising cost of the policy hit a government facing triple-digit inflation and unable to pay its external debts. Further, the price of sugar rose sharply, reducing the supply of sugarcane for ethanol production and thus both reducing the availability of ethanol and raising its price, whilst the price of oil fell sharply, reducing the attractiveness of ethanol. Data presented by Rosillo-Calle and Cortez show that, perhaps surprisingly, sales of ethanol cars held up for another decade, in most years at a level of 150,000 or greater, until the mid-1990s. One factor may have been the full technical support offered for issues such as corrosion and other engine repairs (Sercovich, 1986: 152).

The period 1985–1990 saw Brazil go through considerable upheaval and transition. The year 1985 marked the first civilian government in over twenty years (although not necessarily entirely democratic: Tancredo Neves, who won the elections, died before his inauguration and some felt his nominated vice president, José Sarney, was thus not entitled to take on the role of acting president and, subsequently, president; see also Helfand, 1999). Economic disruption during this period included hyperinflation and the use of four different currencies. There was a debt moratorium in 1987 but, as one senior government official in Brasilia told the authors, Brazil was broke and unable to plan for the future.

This interviewee referred to a huge gap in the energy planning capacity of the government. A major restructuring was needed to deliver greater policy integration. Indeed, oil continued to dominate imports (in 1983, oil represented 57 per cent of all imports; Martins Borges, 1986: 28). That said, a new constitution in 1988 sought to eliminate all permanent subsidies (Nass *et al.*, 2007: 2232). Combined with ethanol supply problems in 1989 (imports being needed from South Africa) and oil prices that remained relatively low, ethanol faced a very challenging start to the 1990s.

With demand for hydrous ethanol falling, there was renewed effort to maintain demand for anhydrous ethanol, via a law signed in 1993 that laid down blending mandates that should be set within the range 20–25 per cent. Given the efforts to expand production, however, new plants were coming on-stream in the early 1990s. That said, given developments in sugar prices, MAPA data show that total ethanol production plateaued through the late 1980s and early 1990s. As one interview in Brasilia revealed, however, a calamitous situation was emerging that was not foreseen at the time: that by the late 1980s, demand for and supply of ethanol were on different trajectories. Thus a combination of low oil prices, high sugar prices, periodic low cane harvests and rising sales of ethanol-only powered cars resulted in growing shortages of the ethanol fuel required. Stories are told of cars just being abandoned because E100 was not available for sale. By 1996, less than 1 per cent of new cars sold were ethanol-powered, down from a peak of 96 per cent in 1995 (Rosillo-Calle and Cortez, 1998: 116).

The political and economic landscape of Brazil shifted in 1990, with the election of President Fernando Collor de Mello. He initiated reforms liberalising the whole economy. The IAA and other bodies had created a complex web of sugar and ethanol market interventions, such that 'Brazil's capacity to switch cane between sugar and ethanol in response to even large changes in oil and sugar prices is greatly limited under current policies' (Borrell and Duncan, 1993: 38). Inevitably, therefore, the sugar industry faced huge changes. The IAA was closed in 1990, Proálcool ended officially in February 1991, and CNAL and CENAL also closed (Hira and Guilherme de Oliveira, 2009: 2454). Some of the functions of the IAA were moved to the Ministry of Industry, then moved on to MAPA. With the abolition of the IAA, exports were liberalised and price controls ended (Sheales *et al.*, 1999: 58).

One senior official interviewed in Brasilia told the authors that, initially, the ending of Proálcool was not expected, given the power of the agricultural lobby. Also, the ethanol industry was sufficiently large that it could not be ignored. Such comments probably reflect the extent to which even well-placed officials, at that time, were not expecting the scale of the reforms actually initiated. That said, during this time the industry was not forgotten totally, notably via the introduction of a mandatory ethanol blend, as noted above. Furthermore, the scale of hydrous ethanol distillation capacity was such that even this part of the ethanol market was fairly secure, given that in such plants, 'alcohol production could not be completely shifted to sugar' (Puppim de Oliveira, 2002: 134).

President Collor remained in post only for two years, before being impeached, but he put in train reforms which were continued by his successors, most notably President Fernando Henrique Cardoso. Under his presidency, in 1997 cane and anhydrous ethanol prices were deregulated, fuel markets liberalised in 1997 and petrol prices deregulated in 1998, with hydrous ethanol prices deregulated in 1999. In 1998, a 40 per cent tariff quota on sugar exports was removed which, along with a large increase in production, plus a 40 per cent devaluation of the *Real*, the currency introduced in 1994, helped also to boost exports (Sheales *et al.*, 1999: 56; see also Martines-Filho *et al.*, 2006; Hira and Guilherme de Oliveira, 2009).

The government, in 1997, also created CIMA, the Interministerial Council of Sugar and Alcohol,[6] whose role initially was to monitor and evaluate the deregulation process as the sector moved to a free market (Martines-Filho *et al.*, 2006: 95). CIMA consists of technical representatives from four ministries (MAPA, Mines and Energy, Finance and Industry Development) who discuss the agenda of the ethanol industry, policy requirements and so on. One senior government official in Brasilia suggested to the authors that this body reflected a particular feature of ethanol policy at the time, that it was more of an agricultural than an energy policy, because of the benefits directed to agriculture, and with ethanol producers lobbying with the Ministry of Agriculture back in the 1980s and 1990s. This followed a trend which saw one factor behind Proálcool being low sugar prices, and that during the 1980s Proálcool was moved into the Ministry of Agriculture (indeed, as we saw earlier, even the establishment of the IAA in 1933 was linked to difficult agricultural conditions for cane growers). As discussed below, however, more recent events have now confirmed ethanol as part of energy policy.

This focus on agricultural support notwithstanding, by the end of the 1990s, government intervention in sugar and ethanol markets was minimal. The most prominent remaining policy was, and remains, the requirement to blend anhydrous ethanol into petrol, with the band still 20–25 per cent. Furthermore, the 'parallel privatisation of many public services, such as transport, utilities and port facilities, [contributed] to reduced costs for Brazil's sugar and alcohol producers' (Sheales *et al.*, 1999: 58). Producers in the NNE region continued to receive an additional R\$5 per tonne of cane delivered for crushing (up to a ceiling of 48 million tonnes). Valdes (2007: 35) notes that producer prices there remain 'significantly higher…than elsewhere'. The subsidy payable to distilleries producing hydrous alcohol was the same for the NNE and CS regions, contributing to a higher percentage of sugarcane in the NNE going to sugar and thence for export. Also influential was the government's decision to assign to the NNE the whole of the (high-priced) preferential access quantity of sugar to the US.

That said, even with all of these reforms the ethanol market in Brazil could not be said to be totally free. The primary problem was not remaining interventionist policies, but a technical constraint on the demand side. The existing car fleet was essentially split into two – cars that could run on petrol blended with anhydrous ethanol up to E25, and cars that could run on hydrous E100. This, in turn, was constraining companies' willingness to produce and supply ethanol of both types, because of rigidities and discontinuities in market prices. For a free market to operate, one more development was needed.

2003 and the introduction of flex-fuel vehicles

In 2003, Volkswagen began selling Flex-Fuel Vehicles (FFVs), capable of running on any mix of (blended) petrol and E100. Thus one car now replaced the former petrol-fuelled car and ethanol-fuelled car. Since the fuel distribution infrastructure already ensured delivery of both types of fuel to petrol stations, drivers were now free to pick up both fuel pumps and fill their cars up with whatever combination was optimal, given the price of each fuel. Previous problems over the uncertain supply of ethanol were also circumvented. If hydrous ethanol supplies fell, the price would rise and consumers would fill up just with E25. If the supply of anhydrous ethanol was affected, to the point of not being able to blend all petrol demanded at 25 per cent, the government could reduce the blend percentage (this is, however, an issue that has also led to some peculiar unforeseen consequences, as discussed in later chapters). That said, as ethanol production in total has risen, the pressure on the government to make such changes has eased.[7]

There were also important psychological benefits for car-owners from this new-found flexibility. Given earlier experiences with ethanol shortages – especially as owners of E100-fuelled cars found it increasingly difficult to obtain fuel – drivers became somewhat sceptical of ethanol as a reliable fuel source. Fuel companies, initially, spoke of FFVs as alcohol-fuelled cars. More successful, it appears, were the advertising campaigns of Volkswagen and the other car-makers that followed, who saw commercial benefits from promoting FFVs on the basis of giving drivers' choice over fuelling options (see also Morceli, 2007).

Managing the new policy environment

Alongside these changes, the structure of policy management changed (as indeed earlier comments indicated it had to). In addition to CIMA, which dealt specifically with ethanol, also established in 1997 was the CNPE (the National Council for Energy Policy).[8] This has a broader perspective for the energy market and is thus led by the Ministry of Mines and Energy (MME).[9] It includes several ministries, NGOs and academics and reports to the president in its role with responsibility for proposing energy policies. MAPA joined in 2008, given its role with biofuels. Interviews with several officials in Brasilia confirmed that this marked the shift to ethanol now being seen as an energy, rather than an agricultural, policy. This was also an important contribution to the efforts to bring greater coherence to the policy process, enabling planning for a more efficient use of Brazil's resources.

Whilst the CNPE reports to the president, concerns were expressed in some quarters during the 2010 election campaign that, if victorious, Dilma Rousseff would try to influence energy policy more than her predecessor, President Luiz Inácio Lula da Silva. When the authors sought to clarify this, interviewees suggested that this was based on either a misunderstanding or misrepresentation of what had occurred previously. As noted earlier, with the abolition of the IAA, responsibilities passed to various ministries. Subsequently (but prior to the election of President Lula), a number of agencies were created, with ministry staff moved across to build up their human and technical capacity. This movement of personnel was matched by planning capacity and policy control also moving to the agencies.

What Mrs Rousseff did, as Energy Minister under President Lula, was to bring back to the ministries the role of energy planning and energy policies (note, for example, the number of ministries on the CNPE), with the agencies in effect implementing the rules and policies established by the ministries and the federal government. Initially, some in the agencies saw this as a reduction in their role rather than a change. Issues did remain, notably the need to ensure the agencies had the appropriate instruments to execute their roles. One example given in interviews concerned Brazil's biodiesel policy, discussed below. Despite the wishes of some, MME had to work within established market rules; for example, it could not give subsidies to the biodiesel industry because they did not have the market instruments to do so.

In short, it was the view of insiders that what Mrs Rousseff did as Energy Minister was not be a hands-on Minister, but essentially reverse the earlier structural changes, outlined above. Thus she returned to the ministries the ability to direct policy, albeit in a more coordinated way than previously. This, everyone interviewed indicated, was not expected to change under President Rousseff. What this discussion indicates is that whilst, in the last decade, there have been quite significant changes to the structure of biofuels policy management in Brazil, this has been done in the context of an essentially free market. There have been occasional time-limited changes made to the petrol–ethanol blend percentage and there are differences in the tax regimes for biofuels and for FFVs, but policy is focused primarily on ensuring markets function efficiently, rather than have a fundamental role in shaping markets, as US and EU policies are seeking to do (as discussed through the rest of this book).

Concluding comments on Brazilian ethanol policy

Government policies in Brazil's ethanol market are now very limited. There remains a blending mandate for anhydrous ethanol in petrol, set typically at 25 per cent but on occasions since 2009 reduced to 20 per cent to accommodate temporary ethanol supply shortages. There are also fiscal incentives in terms of taxes, providing price differentials between petrol and ethanol. Given this and the different performance characteristics of petrol and ethanol, the 'break-even' price is when ethanol is 70 per cent of the price of petrol. The market for ethanol is, however, more or less free – so what enabled Brazil to reach this point?

A crucial economic factor was that the macroeconomic stabilisation efforts saw the government put pressure on the industry to improve competitiveness (a move reinforced by trade liberalisation). Nass *et al.* (2007: 2230) argue that in the 30 years since 1975, ethanol production rose by a factor of 30, yield per hectare increased 60 per cent and production costs fell by 75 per cent. This helped ensure not only the survival but also the expansion of the industry. This, in turn, led to greater critical mass. Partly this arose simply from the length of time Brazil has been producing ethanol. In the context of the main themes of this book, however, there are several policy factors that, by luck, good fortune and good planning, have helped lead Brazil to where it is today.

Governments, including those in the crucial period from 1985 to 1999, established policies which developed and supported, at scale, production facilities for both anhydrous and hydrous ethanol. Moreover, governments have promoted – and consumers accepted – blending mandates in petrol. Serafim (2006: 1) observes that one legacy of Proálcool was an existing fleet of ethanol-fuelled cars, continued production, albeit on a smaller-scale, and the infrastructure to produce and distribute ethanol. During the economic liberalisation of the early 1990s, the government still pushed to reverse the decline in alcohol-fuelled cars, thus maintaining both some level of demand for E100 and a market, in particular, for those distilleries set up as autonomous units to produce hydrous ethanol. Colares (2008: 295) notes that in 1993 the government also raised the ethanol content in petrol, to 22 per cent, a decision which 'effectively saved the Brazilian ethanol program, which would become very useful to Brazil a decade later'. It is interesting to note that as late as 2001, immediately before FFVs were brought to market, the government was once again looking to boost the sale of ethanol-fuelled cars, to promote demand for the hydrous ethanol produced by the installed autonomous distilleries (Puppim de Oliveira, 2002: 136–137).

In a similar vein, one interviewee noted that when economic times were particularly difficult, the government still managed to keep incentives in place at the minimum level necessary to sustain the industry. That said, as Lehtonen (2011: 2430) points out, any difficulties faced in the sugar and ethanol industries will have an absolutely and relatively greater impact in NNE than the CS regions, given that the government's long-standing support for the sugarcane industry in the NNE has had a strong regional developmental dimension to it. As a result, the abolition of the IAA removed the NNE region's 'main defender at national level politics' (Lehtonen, 2011: 2430). Because of the closeness of the region's industry to policy-makers, difficulties had previously been responded to not by entrepreneurialism, but by further policy intervention. Now, 'the more entrepreneurial and the wealthiest among the Northeast sugar elite ... [invested increasingly] ... in sugar and alcohol production in Sao Paulo' (Lehtonen, 2011: 2431). Regional differentials were thus accentuated, as the liberalisation of the Brazilian economy saw the removal of more and more levers of direct policy intervention.

Table 2.1 shows the level – and the volatility – of cane, sugar and ethanol production in Brazil in recent years. The production of sugar has broadly followed the rising production of sugarcane, apart from the period 2007–2008 to 2010–2011, when there was a dramatic rise in the production of hydrous ethanol (for E100 fuel). This also indicates the general absence of a trade-off between ethanol and sugar – given the rise in sugarcane production. Broadly speaking the production of anhydrous ethanol, for blending in petrol, has been stable – albeit with some fluctuations also in this brief period of greater production for hydrous ethanol.[10]

Prodiesel – Proálcool's younger sibling

Biodiesel has long been produced in Brazil, albeit on a very small scale. It was even part of the policy response alongside Proálcool; introduced in 1980, ProOleo sought to promote the use of vegetable oil to substitute for diesel, with a target of a 30 per cent blend (Rosillo-Calle and Cortez, 1998: 115; Pousa *et al.*, 2007: 5394. See also Abreu *et al.*, 2007). Indeed, in 1983 the first patent in the world for the industrial production of biodiesel 'by international norms' was awarded to Professor Expedito José de Sá Parente, of the Federal University of Ceará (having submitted his application in 1980). By the mid-1980s, however, the deteriorating economic situation and falling oil prices contributed to the collapse of the programme. As a result the patent expired after ten years. Another

Table 2.1 Production of sugarcane, sugar and ethanol in Brazil ('000t for cane and sugar; m^3 for ethanol; annual percentage change; and standard deviation of annual percentage changes)

	2000/01	2001/02	2002/03	2003/04	2004/05	2005/06	2006/07	2007/08	2008/09	2009/10	2010/11	2011/12	2012/13	StDev
Cane	2,56,818	2,93,042	3,20,650	3,58,762	3,85,199	3,85,129	4,27,658	4,95,723	5,69,216	6,02,193	6,20,409	5,59,215	5,88,478	
Sugar	16,198	19,218	22,567	24,919	26,685	25,823	29,988	31,026	31,049	32,956	38,006	35,925	38,246	
Total Ethanol	10,592	11,536	12,623	14,736	15,389	15,821	17,844	22,527	27,526	25,691	27,376	22,682	23,226	
Anhydrous	5,621	6,465	7,015	8,876	8,276	8,067	8,030	8,193	9,336	7,065	8,323	8,581	9,844	
Hydrous	4,971	5,071	5,608	5,861	7,113	7,754	9,814	14,333	18,190	18,626	19,053	14,101	13,382	
Cane		14.1	9.4	11.9	7.4	0.0	11.0	15.9	14.8	5.8	3.0	-9.9	5.2	7.3
Sugar		18.6	17.4	10.4	7.1	-3.2	16.1	3.5	0.1	6.1	15.3	-5.5	6.5	8.1
Total Ethanol		8.9	9.4	16.7	4.4	2.8	12.8	26.2	22.2	-6.7	6.6	-17.1	2.4	11.9
Anhydrous		15.0	8.5	26.5	-6.8	-2.5	-0.4	2.0	13.9	-24.3	17.8	3.1	14.7	13.5
Hydrous		2.0	10.6	4.5	21.4	9.0	26.6	46.1	26.9	2.4	2.3	-26.0	-5.1	18.4

Sources: UNICA Data (http://www.unicadata.com.br/historico-de-producao-e-moagem.php?idMn=31&tipoHistorico=2); own calculations.

aspect of this policy failure is highlighted by Stattman *et al.* (2013: 27). They note that unlike Proálcool, where there was strong engagement and support from sugarcane producers, who were facing unstable and uncertain world markets, there were no such factors driving oilseed producers towards ProOleo.

By the end of the 1990s, work was resuming to investigate the potential of biodiesel (Pousa *et al.*, 2007: 5394). In 2002, the Ministry of Science and Technology created a network of institutions tasked with investigating the possibilities for producing biodiesel via the process of transesterification, which involves deriving biodiesel from a combination of oil and ethanol. Interviews with senior government officials in Brasilia revealed that the possibility of a biodiesel policy, incorporating explicit elements of social policy, was first mooted by Agriculture Minister Roberto Rodrigues, in a conversation with President Lula, in January 2003, when a new government took office. He believed biodiesel could successfully be derived from castor oil in the semi-arid areas of the country – including the economically disadvantaged North East. Following positive initial feasibility studies for a biodiesel programme, from July 2003 the programme was discussed in a formal working group that included multiple ministries and other key partners (the outcomes of this process are summarised by Teixeira de Sousa *et al.*, 2008: 128–129).

The policy was finally launched in December 2004, its broad direction coordinated by officials in the Staff Office of the president and managed by the Ministry of Mines and Energy: the National Program for the Production and Use of Biodiesel (PNPB),[11] or Prodiesel. The initial goal was to achieve a B2 blend in diesel that was voluntary at the outset, but made mandatory from 2008. Similarly, there would be a mandatory target of B5 from 2013, with a voluntary increase from 2 per cent to 5 per cent in the interim. Under pressure from the industry, however, the government subsequently brought forward the date of the compulsory 5 per cent blending, to January 2010 (Wilkinson and Herrera, 2010: 757). By 2011, government incentives had created a situation where production capacity was over twice annual output, creating pressure to further increase the blend percentage (Horta Nogueira and Silva Capaz, 2013: 121).

Diesel has, in recent years, represented about half of Brazil's transport fuel needs. Moreover, as noted in Chapter 1, economic growth is expected to result in demand for diesel doubling in ten years (although these estimates predate the recent economic slowdown). Promoting biodiesel thus has clear environmental and emissions benefits. These are compounded by the poor quality of Brazilian diesel, which can be improved by blending with biodiesel. Moreover, whilst Brazil is almost self-sufficient in oil and its derivatives, it is a net importer of diesel. Thus promoting the domestic production of biodiesel also helps enhance energy security.

These issues notwithstanding, its specific design makes Prodiesel unambiguously a social as well as an energy policy (see also Soares Ferreira and Passador, 2010). As one interviewee put it to us, Brazil is looking to use the energy market to promote social development for small-scale agriculture and families in the poorest parts of the country. In addition to general economic conditions, the production conditions of sugarcane and ethanol do not offer the means of taking the North East region forward as happened in the CS. Cane yields in the North East are 57t/Ha, compared with 83t/Ha in the South. Moreover, area and production in the South have doubled in 10 years, but been unchanged in the North East. But, as interviewees pointed out repeatedly, farmers in the North East have strong support in the Parliament and, covering nine states, they represent a powerful voting bloc.

A further dimension to Prodiesel concerns feedstocks. The soybean industry was facing falling prices whilst, at times, the exchange rate has moved strongly against exporters. Interestingly, so*ybeans* are Brazil's largest farm export – not the processed meal and oil. Before Prodiesel, soybeans typically were exported to China and Argentina, where the beans were crushed to extract the oil and meal. Prodiesel has encouraged a crushing industry to set up in Brazil. There have been wider implications for soybean producers as well. In Mato Grosso State, the area of greatest soybean production in Brazil, production costs are low but poor infrastructure meant that it was difficult and expensive for producers to get their output to ports for export. Prodiesel has seen not only mills setting up in Mato Grosso; new ports were being built nearer to the main centres of production. Despite this, not only was castor oil the feedstock of choice for Minister Rodrigues, as it was well-suited to the semi-arid climatic conditions of the region, it was initially also the industry's main choice (Pousa *et al.*, 2007: 5395. Abreu *et al.* (2007) discuss the issue of feedstocks in detail, region by region).

Despite the social policy goals and wider economic gains to be had, the government faced a significant challenge: unlike Proálcool, Prodiesel was introduced into a liberalised economy. This limited significantly the policy levers available to it. As a result, the tax system was an important instrument, in particular in terms of giving incentives to the producers of biodiesel (see Turba and Fajardo, 2008: 105–106 for details). Alongside this, the Social Fuel Certificate (SFC) was developed. Qualification for the SFC varies by region, given the varying nature of small-scale agriculture in different parts of the country. Taxes vary by region, by raw material produced and by the type of agriculture producing the feedstock. For biodiesel producers to qualify for the SFC and thus gain the tax benefits, they must do three main things:

- Buy a minimum quantity of raw material from small farmers.
- Sign a contract with each family, supported and co-signed by a representative of a rural union recognised by the Ministry of Agrarian Development (MDA), which must include remuneration for a minimum of 12 months.
- Provide families with technical assistance, certified seeds, equipment, access to fertilisers, advice on good practice, etc. This can include the hiring of agronomists.

The MDA then issues the SFC. It also has auditors who check to ensure the biodiesel producers are complying with these obligations. Biodiesel producers have different quotas for each region specifying how much they must buy from small farmers. The biodiesel auctions (see below) are separated into two markets, where 80 per cent of demand is reserved for holders of SFCs, and the remaining 20 per cent can be bid for by any biodiesel producer. Officials in MME informed us that, by late 2010, 93 per cent of biodiesel was coming from SFCs; but this does not mean 93 per cent of raw material comes from small farmers. Rather, 93 per cent of biodiesel has, in its content, a share coming from small agriculture. The process is also a slowly developing one. The government faces a challenge to restructure rural areas and communities through agrarian reform. MDA has the challenge to integrate small farmers at 'the proper time' (to quote one interviewee in Brasilia) and so is phasing-in the process. In particular, after initial contracts have been awarded, it can then take two or three harvests before a longer-term contract is awarded.

Partly, this slow process was influenced by events in the early years. A biodiesel company in the North East established a contract but because of a very aggressive commercial strategy (in seeking to become Brazil's largest producer by 2007) almost went bankrupt as a result. Combined with the commodity price spike in 2008, where biodiesel producers were unable to fulfil the contracted biodiesel prices, family farmers started to look negatively on the programme. The MDA thus recalibrated the pace of SFC issuance to protect vulnerable small farmers: it is better to go slow but consistently. This message was repeated several times to the authors by government officials in Brasilia.

Another consequence of the high fuel prices in the early years of Prodiesel was that the tax incentives were insufficient to deliver the desired volumes of biodiesel on a consistent basis. This was constrained still further because, given the federal nature of Brazilian tax policy, the point was reached where national tax levers were being compromised by state taxes, as the government sought to induce further increases in biodiesel supply. As a result, MME and the CNPE established biodiesel auctions; and they made mandatory the blending of biodiesel produced with the SFC. In 2006, all biodiesel produced under SFCs had to be blended. Although the intention was to have the auctions only until the end of 2008, thereafter leaving it to the market to drive supply and demand, biodiesel producers and fuel distributors asked the government to keep the auctions in place: there was concern that by asking for the blending of a higher-priced product (biodiesel) with lower-priced diesel, fraud could result.

With the auctions in place, the industry expanded to the point where the move to mandatory B5 could safely be brought forward, as indicated earlier, to B5 in January 2010, with interim mandates of B2 in January 2008, B3 in January 2009 and B4 in July 2009. Moreover whilst in 2006, when the auctions were designed, the installed capacity was insufficient to deliver a mandatory B2 blend, by 2010 it was sufficient to deliver B10 – something the private sector had been asking for but, for the reasons set out above, the government had so far resisted. Yet another reason for taking Prodiesel slowly was that, unlike Proálcool, Prodiesel did not have an established infrastructure network to tap straight into (Teixeira de Sousa *et al.*, 2008). Rapid expansion of production could, therefore, result in supply bottlenecks.

Interviewees gave the impression, rather than stating explicitly, that one reason for the social dimension to Prodiesel was the poor social performance of Proálcool. Thus another reason for developing Prodiesel slowly was to allow time for the design of a system which embraced small farmers. Even today, this remains the biggest challenge for the policy. Another, related, concern and reason for the steady pace is the desire to change the balance of feedstocks used. In the early years, the mix was roughly 80 per cent soybeans, 20 per cent animal fats, but there is a policy, including an agro-ecological zoning (AEZ) scheme, to boost palm oil production and use.

It was suggested by some interviewees that soybean production and use would be dominated by large producers without SFCs. Whilst unable to compete in a free market, SFCs can help small producers especially given that, in some states, half of the soy produced comes from small-scale family-farmed agriculture, which is able to deliver good yield and quality. According to Horta Nogueira and Silva Capaz (2013: 120), however, most biodiesel production in the early years was focused on the Centre West, with its established agri-industrial infrastructure. Wilkinson and Herrera (2010) paint a gloomy picture for small-scale farmers, suggesting that their continued involvement with biodiesel is dependent on government policies continuing.

Studies looking at the impact of Prodiesel indicate that whilst it is having a positive economic impact, the scale is much less-so than that originally desired by policy-makers. That said, there is also agreement that the policy represents a sound basis for continuing to pursue the goals originally set for the policy, albeit with necessary changes and developments to key elements. The view of Schaffel *et al.* (2012: 733), based on a regional case study of castor in Bahia state, North East Brazil, is that whilst the social certificate process 'remains an essential strategy for future sectorial development, particularly the integration of poor family farmers', sustainable development requires a broader approach than just this. They point out that certification as a means of engagement with small, poor farmers focuses primarily on economic sustainability. This may indeed help to address sectoral development but, of itself, it does not address social or environmental concerns. Even its effectiveness in promoting economic sustainability has been attenuated by relatively low participation rates, production inefficiencies and, in the case of castor, farmers who operate on a small scale, planting castor only as a minority crop alongside subsistence food crops. On the plus side, one channel for potential gains is the technical support for small farmers that comes as a pre-requisite for companies' participation in the scheme.

Rathmann *et al.* (2012: 98) also point out problems of the inefficiencies of small-scale family farmers. Only a minority of Brazil's biodiesel refineries are located in the NE region – and 'these plants have large levels of idle capacity because of the inability of family farmers to grow sufficient oilseeds to supply their needs'. These refineries are having to rely on large-scale agribusinesses to supply the required feedstocks, contrary to one of the principal goals of the policy. These authors highlight another problem with using soybeans as a feedstock – that soya-based biodiesel is inferior technically to mineral diesel. This reinforces the need to diversify feedstocks, notwithstanding the challenges noted by Schaffel *et al.*, and others. César and Batalha (2013: 173), for example, make an important general point in their analysis of the potential of palm oil – that 'the cost of implementing projects with family farmers is higher in deprived areas of difficult access, and especially in regions with no tradition in cooperativism and large scale production'. They indicate this is the case with palm oil, but it probably applies more generally. This reinforces the argument of Schaffel *et al.*, that broader policy instruments are needed, in addition to the Social Fuel Certificate.

Given the problems experienced in the early years of Prodiesel, insofar as its goal of social policy goes, it appears that policy reform, at the very least, would be needed. A continuation of some sort of biodiesel policy intervention on behalf of small-scale farmers that will actually deliver not only on the economic, but also the wider social goals of the policy really will, it appears, test the political influence of the representatives of this part of Brazil.

As a final observation on Prodiesel, it is important to note that it was conceived as a domestic policy. Indeed, at the time there was no international market for biodiesel, whilst the international market for ethanol only really began in 2005, with the US Energy Policy Act (see Chapter 4). This is critical to understanding what Brazilian policy-makers mean when they refer to the sustainability of Prodiesel. This term is meant in the context of the emerging *domestic* market and, with the SFCs, refers to social as well as economic sustainability. It was also pointed out to the authors that this includes a consideration of access to labour, as Brazil's birth rate is falling, and school participation rates are rising.

Conclusions

Brazilian policy stands in contrast to both cases to come – and we offer it as a point of comparison as we analyse EU and US policy. The ethanol industry in Brazil has, in terms of capacity to produce on a large scale, had a much longer history than the industries in the EU and US. Whilst, as we shall see in Chapter 4, ethanol was a much-vaunted fuel in the early years of the motor industry, the availability of cheap domestically sourced oil saw ethanol decline rapidly.

Until recently, Brazil lacked these domestic energy resources. What it did have, however, was considerable resources to develop renewable energies – of which ethanol was just one component. As regards transport fuel, this domestic capability came into its own in the 1970s, in the wake of the first oil crisis. The technical know-how to produce both sugarcane and ethanol, the infrastructure and extensive land area on which to grow the sugarcane enabled Brazil to address via ethanol what, in 1973 and 1974, had become a major economic challenge – its dependence on imported oil with price rises putting enormous strain on the economy. That the spike in sugar prices was so short-lived, such that very quickly cane growers were facing very low prices for sugar, led to pressure for an ethanol policy from agricultural interests as well.

Over time, however, ethanol has played a major role in the Brazil's global re-positioning, as a leader in renewable energies. The environmental performance of Brazilian sugarcane ethanol is widely regarded, on multiple dimensions, as the best in the world of all current first generation biofuels. It does, however, contribute more than just ethanol, given the use of sugarcane in electricity generation as well (for a detailed multidimensional analysis of this, see CGEE, 2012). The more recent biodiesel policy plays a somewhat different role in the economy. It is partly a response to the projected growth in demand for diesel in the coming years, driven by economic growth. It is also being used as a socio-economic policy, whereby a portion of the biodiesel market is intended to be reserved for small-scale farmers.

One of the most significant features of Brazil's biofuel market – and the starkest point of contrast with both EU and US positions currently – rests with ethanol and the fact that it is, essentially, a free market. The key development in this regard, of flex-fuel vehicles, in 2003 has given freedom to both the demand and supply sides of the market. As we shall see in Part II, this is not without its problems, but it does mean that the nature of government interventions in biofuels markets in Brazil, but especially of ethanol, is very different to those of the EU and US that we analyse in Chapters 3 and 4. Related to this, interviews confirmed that the second key dimension of Brazilian policy, alongside FFVs, is the increasingly coordinated and coherent medium- and long-term planning capacity, the development of which was set out earlier. As we shall see in subsequent chapters, policy capacity is something which sets Brazil's policy apart.

3
EU Biofuels Policy – An Introduction and Overview

In this chapter we review the development of EU biofuels policy. We shall see clear evidence of all three of the main biofuels policy drivers at work, but with the environmental-climate change driver having a bigger role in the genesis of current EU policy than is the case with US policy. Moreover, as demonstrated in Chapter 2, whilst the environmental dimension is now a central aspect of policy in Brazil, it was not always so. As with Brazilian and US policy, clear distinctions must be drawn between supply and demand of biofuels, and between ethanol and biodiesel. In the EU, however, an additional dimension to incorporate is that between the member states and the (supranational) EU. As will be seen, policy was initially under national jurisdiction, with common EU policies emerging only relatively recently, yet rapidly. Nevertheless, with many elements of policy enacted by Directive rather than Regulation, there still remains considerable national variation in implementation, as elaborated on below.

In analysing the origins and early development of EU biofuels policy, a further distinction can be drawn between biofuels policy as an element of renewable energy policy and broader climate change policy. As will be seen, these distinct strands of EU policy activity have circled around each other, coming together only in 2008–2009.

The emergence of the EU as an international climate policy actor

The late 1980s to early 1990s was a period of significant change for the EU. The Single European Act (SEA), signed in 1986, created the Single European Market (SEM) and, with it, a more unified economic identity for the EU. The SEA also introduced environmental policy into the

Treaty of Rome from 1987. The Maastricht Treaty,[1] signed in 1992, set out the course to the introduction of the euro as a common currency, whilst the SEA and Maastricht Treaty extended Qualified Majority Voting to more policy areas, including environmental measures. The fall of the Berlin Wall led to German Unification and precipitated the move towards the EU enlargements of 2004 and 2007. Domestically, ever more EU member states saw the rise of Green Parties through the 1970s and 1980s, to the point that by the late 1990s, Green Party politicians sat not only in the European Parliament, but also in 11 out of the 15 national parliaments (Kelemen, 2010: 340). This development triggered greater environmental policy activism in the member states, to the point that this legislation threatened the SEM, as a non-tariff barrier to free trade (Kelemen, 2010: 340).

This confluence of factors also influenced the wider context of EU policy-making. Sbragia (2005) talks of this in terms of institution-building, whilst Vogler (2011) analyses the emergence of EU policy 'actorness'. In short, this period sees the growing stature of the EU, both alongside the member states (through the sharing of policy competences in EU policy-making) and globally, recognised increasingly as a collective, unitary policy actor in its own right. Given the national political shifts outlined above, environmental and climate change policy offered an ideal arena for the EU to take on the mantle of global policy leader. This push was led by the European Commission and has involved, increasingly, the European Parliament. The reversal of decades of US policy leadership, in particular under President George Bush Sr. (from 'global environmental leader' to 'laggard and obstructionist', as Kelemen, 2010: 336, puts it), created a vacancy that the EU was more than willing to fill.[2]

The first major event at which the EU took up its newly adopted role, 'as an equal participant' (Kelemen, 2010: 340), was at the Rio 'Earth Summit' of 1992.[3] This summit saw an international treaty negotiated, the United Nations Framework Convention on Climate Change (UNFCCC), although this treaty contains no legally binding emissions reduction targets. When the Conference of the Parties to the UNFCCC had their first annual meeting in 1995 ('COP 1'), they agreed to work towards clear, firm commitments, a negotiation process that was concluded two years later, at COP 3 in Japan, with the signing of the Kyoto Protocol (Barnes, 2011: 47), although it did not enter into force until February 2005.

Sbragia (2005: 215–216) argues that whilst the member states played the leading role in much of the negotiating prior to Rio, especially during bilateral talks with the US, not only did the member states adopt a clear unified position, it was already being recognised that there was no clear separation of national and EU interests or positions. In the period from 2005 to 2007 the EU, as a single actor, played a leading role. The agreed baseline against which the agreed CO_2 emissions would be cut (1990) was, undoubtedly, advantageous to the EU. First, the 'dash to gas' in the UK, and the closure of inefficient and polluting factories in eastern Germany following unification, meant substantial emissions cuts had already occurred relative to the base year. Second, there was a positive externality from this, given the burden-sharing approach agreed by EU member states which saw those cuts shared across the member states. Third, with domestic political changes putting greater focus on emissions, a multilateral negotiating context would allow the EU to try to leverage emissions-reductions across more countries (Kelemen, 2010: 343–345).

The EU showed clear and strong leadership, agreeing the largest emissions cut of all developed countries, whilst playing a major role in the negotiations, ensuring some of the more reluctant participants, such as Russia, also sign the protocol (see the aforementioned references for more details). An interesting feature of EU policy at this time, however, is that despite these international successes, the EU faced a credibility gap, given the lack of matching domestic policy developments (Oberthür and Pallemaerts, 2010: 36). That said, as we shall see below, this situation was to change dramatically in the next decade. Indeed, one interpretation of this is that international agreement was used as a form of policy pre-commitment domestically.

After a very brief review of early biofuels-related developments, in the EU member states individually, then collectively through the EU, we shall see how this subsequent policy development took place. It is worth noting at this stage, however, that whilst this international leadership provided the context within which the establishment of biofuels policy took place, more recent analyses have led to this leadership role being questioned. Skovgaard (2013) talks about more recalcitrant EU member states being trapped by circumstances in the period prior, in particular, to the COP15 negotiations in 2009. Subsequently, the factors explaining this entrapment were diminished or absent which in turn – Skovgaard argues – explains the inability of the EU members to agree to additional emissions cuts, notably by raising the emissions reduction from 20 per cent to 30 per cent. A key factor was the increasing engagement of economic interests in environmental negotiations post-2009.

The emergence of an EU biofuels policy

As we saw in Chapter 1, the production and use of biofuels began in the 19th century. Kutas *et al.* (2007: 15) report that France was in the vanguard of the use of denatured alcohol as a fuel at that time. Later, between 1920 and 1950 there was 'significant' substitution using ethanol produced from beet. Early developments focused on ethanol, as the use of diesel engines in cars was extremely limited. By the 1960s, however, a combination of abundant cheap oil and a lack of surplus sugarbeet saw even the production and use of ethanol largely disappear.

By the 1980s, reflecting shifts in domestic politics, the EU began to promote renewable energy. The focus of Council Directive 85/536/EEC (5 December) was 'on crude-oil savings through the use of substitute fuel components in petrol', having regard, as the Preamble put it, to 'the present energy situation'. Such a move would reduce the EU 'dependence upon imported crude oil' and broaden 'the raw materials base' for transport fuel. In short, this early Directive on promoting renewable energy was motivated by energy security concerns. In a manner seen again in the 2000s, the measures implemented focused primarily on the supply side, via agricultural policy, along with 'eventually, targets for biofuel consumption' (Kutas *et al.*, 2007: 15). Thus 'The first policies to benefit biodiesel production in Europe were not specific to biodiesel' (Kutas *et al.*, 2007: 15), but came instead through the Common Agricultural Policy (CAP). There were minimum price guarantees or area payments for 'feedstocks [, which] comprise the greatest operating cost for biofuels'. The set-aside policy, requiring large farmers to remove a certain proportion of their arable land from production as a means of reducing surpluses, permitted them to produce crops for non-food purposes on set-aside land, 'for example oilseed rape for biodiesel' (Kutas *et al.*, 2007: 15), and still be eligible for the set-aside payments (see Ackrill, 2000: 65–69, for more on set aside).

From 1992, countries could offer tax exemptions for pilot projects developing 'fuels from renewable resources'.[4] Also, in the early 1990s some EU countries began to develop quality standards for biodiesel, creating greater product and market certainty (Kutas *et al.*, 2007). Meanwhile, the 1995 EU accession of Austria, Finland and Sweden reinforced the critical mass of member states advocating environmental and climate-change policies. Thus, by the late 1990s, the expansion of biofuels production in a number of member states led to growing calls for more coordinated EU policy action. The 1985 Directive focused on petrol, but later policy actions were aimed more at biodiesel. The EU was a significant importer of diesel but an exporter of petrol, thus the policy shift helped correct a glaring policy error, where energy security was claimed for a policy prioritising petrol, when fuel imports were dominated by diesel. As a result of these measures, by 2004 80 per cent of EU-produced biofuels were biodiesel (Schnepf, 2006).

By the mid-1990s, a range of EU documents emerged which addressed renewable energy generally, locating biofuels within a broader climate change narrative. In so doing, this pre-empted by over a decade the broad policy context within which the 2008/2009 biofuels legislation was located, as discussed below. A 1994 proposal for a 'Directive on excise duties on motor fuels from agricultural sources'[5] refers, in the Preamble, to both environmental factors and the potential for rural development gains, by creating a new market for agricultural commodities, and reducing (rural) unemployment. The rural development dimension also appears in the 1995 Energy White Paper 'An Energy Policy for the European Union',[6] whilst a scenario analysis of 'European Energy to 2020' models the extent to which biofuels could potentially play a part in road transport fuel by 2020.[7]

These and other documents of this period are discussed in a key report, 'Energy for the Future: Renewable Sources of Energy'.[8] This 'White Paper for a Community Strategy and Action Plan' identifies how the EU can progress towards the target that, the following month, was to be presented to the Kyoto climate change negotiations: by 2010, greenhouse gas (GHG) emissions should be reduced by 15 per cent relative to 1990 levels. That said, the Commission argues on page 5 of the White Paper that action is needed on renewable energy 'whatever the precise outcome of the Kyoto Conference'. In the end, the outcome was a commitment to reduce GHG emission by 8 per cent, the largest cut committed to by developed countries.[9]

The 1997 White Paper, as already noted, had climate change concerns as a key driver, whilst also referring throughout to energy security and job creation benefits, including in rural areas. Given how EU policy was to develop a decade later, it is interesting to note that 'The Committee on Agriculture and Rural Development of the Parliament has also issued an Opinion in which it considers that the contribution of biomass-derived energy to the primary energy mix could reach 10% by 2010' (European Commission, 1997: 9). The Commission meanwhile, talked about reaching 12 per cent renewable energy penetration by 2010. 'As far as the market sectors are concerned, the doubling of the current electricity and heat production from renewables plus a significant increase of biofuel in transport fuel use by 2010 are important elements in the scenario for achieving the overall Union objective'. (European Commission, 1997: 11). This statement is a crucial staging post in the development of EU policy, as it recognised renewable energy would have to penetrate all three energy markets – power, heat and transport fuel – with explicit reference to a greatly expanded role for biofuels in the transport fuel mix.

In 2000, a proposal was made for measures to promote the generation of electricity from renewable sources. Specifically, electricity generated from renewable sources should contribute 22.1 per cent towards the overall White Paper target of 12 per cent of gross inland energy consumption coming from renewable sources by 2010. The resulting Directive entered into force in October 2001.[10] Also in 2000, at the Lisbon Summit, the goal was set for the EU 'to become the most competitive and dynamic knowledge-based economy in the world capable of sustainable economic growth with more and better jobs and greater social cohesion'.[11] In Gothenburg the following June, the European Council 'added an environmental dimension to the Lisbon process for employment, economic reform and social cohesion' (Paragraph 1 of the Presidency Conclusions)[12] with the aim of 'decoupling economic growth from resource use' (paragraph 21). Interestingly, the Presidency Conclusions from the Helsinki Summit of December 1999, in paragraph 50, invite the Commission 'to prepare a proposal for a long-term strategy dovetailing policies for economically, socially and ecologically sustainable development to be presented to the European Council in June 2001. This strategy will also serve as a Community input for the ten year review of the Rio Process scheduled for 2002'. It would thus appear that the intention all along was to include a sustainability dimension into what became the Lisbon Strategy, but it was also intended, ab initio, to be part of the EU's wider, global, environmental engagement.

Prior to the Gothenburg Summit, the Commission had prepared a Communication, to be presented to the Summit (Commission of the European Communities, 2001a). On page 2, it is recognised 'that in the long term, **economic growth, social cohesion and environmental protection must go hand in hand**' (emphasis in original). Then, on page 10, the target was laid down that 'Alternative fuels, including biofuels, should account for at least 7% of fuel consumption of cars and trucks by 2010, and at least 20% by 2020. The Commission will make a proposal in 2001 for adoption in 2002'. This proposal (Commission of the European Communities, 2001b) was made less than a month after the Directive on renewable electricity was adopted.

The three policy drivers introduced in Chapter 1 – energy security, rural development and GHG emissions reductions – are identified explicitly as goals to which biofuels are intended to contribute (Commission of the European Communities, 2001b: 37). In the Explanatory Memorandum introducing the proposed legislation, however, rural development and rural employment are identified separately, under 'Impact on Other Policies' (see Commission of the European Communities, 2001b: 21–24). Interestingly, this chimes with a comment to the authors from a senior Commission official who played a central role in drafting the proposals for the key piece of EU legislation, the Renewable Energy Directive (RED), analysed in detail below. Referring specifically to the RED, he was very clear about the causality in terms of rural employment: it was great that the EU could point to the fact that its biofuels policy would create rural jobs, but the creation of rural jobs was not a key driver of biofuels policy. It could, it was felt, stand on its own (climate change-mitigating and energy security promoting) feet.

Although the 2002 European Council target date for legislative agreement was missed, agreement was reached in May 2003 on what is often referred to simply as the Biofuels Directive (EU, 2003). This Directive can be taken as the starting point of EU biofuels policy although, as we have seen, the member states had been promoting biofuels for some years (see van Thuijl and Deurwaarder, 2006, for a summary; Bomb *et al.*, 2007, for further details on German and UK policy). Comparing the titles of Commission of the European Communities, 2001a and 2001b, we can see that the Biofuels Directive has particularly strong ancestry in the broader sustainability agenda. Bearing that in mind, and with an eye on future policy developments to come, it is worth exploring in some detail the amended proposal presented by the Commission in September 2002, following amendments adopted by the European Parliament (Commission of the European Communities, 2002), some months before the Biofuels Directive was adopted.

Several of the Parliament amendments addressed the environmental aspects of biofuels. A new paragraph 25, introduced into the opening Recital and retained in the Directive, stated that 'An increase in the use of biofuels should be accompanied by a detailed analysis of the environmental, economic and social impact in order to decide whether it is advisable to increase the proportion of biofuels in relation to conventional fuels.' Paragraph 24, also retained in the Directive, argued that 'Research and technological development in the field of the sustainability of biofuels should be promoted.' On the other hand – and significant for later discussions – part of the amended proposal, regarding the text of paragraph 15 of the opening Recital, did not appear in the final Directive. This stated that 'In order to ensure sustainable farming practices, a set of clear environmental criteria for the production of liquid biofuels must be established'. Moreover, the amended proposal also removed those elements of the original proposal that would have had the result of creating a minimum *mandate* for the blending of biofuels into transport fuels.

From the biofuels directive to the renewable energy directive[13]

The Biofuels Directive set indicative (as opposed to mandatory) targets, for the incorporation of biofuels and other renewable fuels in the EU transport fuel mix, of 2 per cent (by energy content) by the end of 2005, and 5.75 per cent by the end of 2010. In addition, revised legislation was agreed, in the form of the Energy Taxation Directive, which updated the 1992 measures allowing for fiscal incentives to be offered for biofuels production.[14] The Biofuels Directive required the member states to report annually on the measures taken to promote biofuels production and use, the annual indicative targets set and the extent to which progress towards the targets was being made. In addition, no later than the end of 2006 the Commission was to produce a report covering a range of issues. These included, inter alia, the economic and environmental impacts of increasing the use of biofuels and other renewables, a life-cycle analysis to determine those renewables that deliver on the climate goal whilst also being competitive, the sustainability of land use for growing biofuels feedstocks and – if necessary – considering the introduction of mandatory biofuels targets in order to ensure delivery on the policy goals.

Over the period 2004–2007 the EU, led by the Commission, undertook a great deal of work, both reflecting on policy successes and failures to date and analysing future possibilities for renewable energy in heating, power and transport fuel. Picking up here just some of the key threads from this flurry of activity, the greatest progress had been made in (electricity) power generation, less progress was made in biofuels and less still was made in the area of heating (and cooling), where no EU legislation at all had been brought forward.

In 2004 the Commission put it bluntly when it said the main cause of the likely failure to hit the 2010 target of 12 per cent of energy coming from renewable sources was down to 'sluggish growth of renewable energy markets for heating and cooling' (Commission of the European

Communities, 2004: 5). This document, it is important to note, is indicative of many of the Commission documents published in this period, in that it locates policy efforts on biofuels firmly within the wider policy issue of renewable energy promotion. Another interesting dimension is the location of biofuels within a parallel debate over biomass. This was recognised as a source of renewable energy which could contribute to heating, power and transport fuels. Another parallel debate was that of improving energy efficiency, which would help limit demand growth across all three dimensions of energy use (see, for example, Commission of the European Communities, 2005a).

December 2005 saw the Commission push for progress on renewable energy, publishing Communications dealing with the promotion of electricity generation from renewables (Commission of the European Communities, 2005b) and setting out a 'Biomass Action Plan' (Commission of the European Communities, 2005c). Soon after, it published 'An EU Strategy for Biofuels' (Commission of the European Communities, 2006a) and, a month later, a Green Paper on renewable energy which developed aspects of the revamped Lisbon Strategy (Commission of the European Communities, 2006b). Late 2006/early 2007 saw more publications, including a progress report on the production and use of biofuels in road transport (Commission of the European Communities, 2007a) and a Communication setting out a 'Renewable Energy Road Map' (Commission of the European Communities, 2007b).

All of this activity was leading up to proposals, presented in January 2008, for a set of legislative measures which would mark a step-change in EU renewable energy policy. They constituted, collectively, the Climate and Energy Package. This included two Directives key to developing biofuels in the EU – the RED (EU, 2009a) and a revised Fuel Quality Directive or FQD (EU, 2009b). These, supported by the European Parliament in December 2008, were finalised by the Council in April 2009 and enacted two months later.[15]

Interviews with policy insiders in Brussels revealed how much work had to go into negotiating and agreeing so much legislation in (relatively) so little time. The aim was to have everything agreed and in force before the COP15 meeting in Copenhagen in December 2009 where, as a result, the EU could 'demonstrate global leadership for a long-term international climate regime' (Kretschmer *et al.*, 2009: S285). Some, however, also wished to get political agreement on the legislation before the end of 2008 because the Czech Republic was to take over the rotating Presidency of the Council in the first half of 2009 and the Czech president at the time, Václav Klaus, is a climate change sceptic. The fact that the legislation was delivered on time can, in part, be explained by the 'mutually reinforcing partnership' between the European Council and European Commission (Bocquillon and Dobbels, 2014: 30). This occurs where 'the European Council and Commission can complement and strengthen each other's initiatives' (Bocquillon and Dobbels, 2014: 27).

The Climate and Energy Package established the so-called 20–20–20 targets, which in turn formed one part of 'Europe 2020' programme for smart, sustainable and inclusive growth, the successor policy to the Lisbon Strategy, that by 2020:

- there should be a 20 per cent reduction in EU GHG emissions, relative to 1990
- there should be a 20 per cent improvement in EU energy efficiency, relative to 1990
- 20 per cent of EU gross final energy consumption should come from renewable sources

With the third element, each member state has its own national target, determined with reference to existing performance. That said, within this figure there is a supplementary target that is applied uniformly to all member states, that by 2020:

- 10 per cent of transport fuels will come from renewable sources.

The flat-rate application of this figure across all member states is because fuel technologies in general, and biofuels in particular, are easily tradable within the SEM. EU biofuels policy is now firmly embedded in wider EU environmental and energy policy. It is linked to climate change mitigation, but also to economic growth (through Europe 2020), the SEM (given the continued failure to deliver an integrated internal market across the EU for gas and electricity) and wider energy security concerns.

The earlier discussion outlined briefly the trajectory of the EU debate towards the inclusion of biofuels in the Climate and Energy Package. The debate that led to the establishment of the 20–20–20 figures (and the 10 per cent figure for transport fuels) is itself worthy of discussion. By January 2004, studies had indicated a 20 per cent target for renewable energy in total energy consumption by 2020 was feasible (reported in Commission of the European Communities, 2004: 6). It was recognised even at this stage, however, that the absence of a policy promoting renewables in heating was a problem, whilst the measures supporting the (non-mandatory) targets for electricity and biofuels were inadequate to deliver even on the targets set for 2010, let alone this more ambitious figure for 2020. Other studies had, similarly, confirmed the feasibility of a target for a 20 per cent improvement in energy efficiency (Commission of the European Communities, 2005a: 16). Running through these documents is also a strong message that, if emissions reduction targets are to be achieved, policy instruments need to be coordinated and reinforced.

Work analysing the potential for biofuels began even earlier. In 1996 a European Commission modelling exercise estimated a 'Conventional Wisdom' or Business As Usual share of biofuels in the transport fuel fleet of 9.6 per cent. Were countries to collaborate and adopt more radical pro-environmental taxation, the figure rises to 16 per cent.[16] By 2001, the European Commission was arguing that biofuels should make up at least 7 per cent of fuel by 2010 and at least 20 per cent by 2020 (Commission of the European Communities, 2001a: 10). This latter figure was repeated in the Preamble to the 2003 Biofuels Directive (Recital paragraph 17) but did not appear in the legislation, which set only indicative targets for 2005 and 2010. On a similar trajectory the European Council, at their March 2006 Summit, proposed targets for 2015, of 15 per cent renewables in total energy consumption, with an 8 per cent target for biofuels.[17]

By the end of 2006, a firm proposal of 10 per cent biofuels in transport fuel by 2020 was being called for (Commission of the European Communities, 2007b: 10), whilst modelling exercises analysing the impact of expanded biofuels production, primarily through domestic means, were undertaken assuming a 14 per cent share in 2020 (reported in Commission of the European Communities, 2007a, notably through pages 9–11). The European Parliament was in favour of the higher figure for biofuels, but sought a higher figure still, 25 per cent, for energy coming from renewable sources by 2020 (reported in Commission of the European Communities, 2007b: 3).

The 'Biofuels Progress Report' (Commission of the European Communities, 2007a) called for a 10 per cent target to be set. The accompanying 'Review of Economic and Environmental Data' (Commission of the European Communities, 2007d) modelled 7 per cent and 14 per cent incorporation rates. Its findings were summarised in the 'Impact Assessment to the Renewable Energy Road Map' as indicating 'that – on present knowledge – the optimum share of biofuels in 2020, within an overall renewable energy target of 20%, will be in the region of 14%. However, in fixing a minimum target, and one which should be binding, a more cautious approach should be adopted'. (Commission of the European Communities, 2007c: 26). Moreover, it makes clear that

14 per cent would only be optimal if certain conditions are met. These include ensuring imported biodiesel was based on sustainably produced feedstocks, and recognising the possible impact on food prices.

Our interviews in Brussels indicated that, within the Commission, the Agriculture Directorate General had pushed for the lower 10 per cent figure rather than 14 per cent, on the basis of the aforementioned concerns. NGOs, however, were expressing concern that even the 10 per cent figure was sufficient to generate negative effects – including damaging environmental impacts. Moreover the European Council, even in 2007, felt the 10 per cent figure should be conditional: 'The binding character of this target is appropriate subject to production being sustainable, second-generation biofuels becoming commercially available and the Fuel Quality Directive being amended accordingly to allow for adequate levels of blending'.[18] These qualifications were subsequently incorporated in the Preamble to the RED. Whilst the FQD was amended and sustainability criteria introduced, only limited amounts of second generation biofuels have been brought to market. We return to this point below, when we analyse a proposal, tabled in late 2012, for reform of the biofuels-related legislation.

The year 2007 was a key point in the development of what would become the Climate and Energy Package, with its memorable (if suspiciously convenient) targets. At their March 2007 Summit in Brussels, the European Council laid down an Action Plan for 2007–2009 – an 'Energy Policy for Europe (EPE)'.[19] This document did not, at this stage, lay down the 20–20–20 targets, but re-stated what had appeared in previous documents and analyses. There was also a growing acceptance that, if targets were to be reached (whatever those targets were), they would have to be mandatory. Indeed, by 2007, national policies were already including mandates, alongside or instead of tax-based policies (Commission of the European Communities, 2007a: 7). Mandates offer a lower-cost and more robust way of delivering on the target, whilst also offering investors greater certainty about the future trajectory of the market.

That said, our interviews conducted with officials in Brussels indicated that several countries, including Germany, were concerned by mandates. With Germany holding the Presidency of the European Council in the first half of 2007, a compromise deal was struck. Referring to the third policy target above, the goal was for a single, aggregate target, of 20 per cent share for renewables in final energy consumption by 2020. Thus far, discussions had been based around having separate targets for renewable energy usage in transport, heating and power. Mandates were also supported by producers' organisations, as they would require the strengthening of policy efforts to develop new markets for agricultural feedstocks.

Why, though, was a separate figure of 10 per cent set for transport fuels, within the overarching 20 per cent goal? One reason was, as set out in Chapter 1, the need to address a sector expected to be the single largest source of emissions growth in the coming years. Beyond this, interviews in Brussels revealed it was believed that whilst the 20 per cent figure would give investor certainty for power and heating, it would not, by itself, send a clear signal to investors in transport fuels; nor would it give any indication about the likely scale of action that was felt was needed in order to deliver on GHG emissions reduction goals. The quid pro quo for this separate target for transport fuels was that it would make no distinction between ethanol and biodiesel, nor between different feedstocks, nor between different biofuel generations. Indeed, in contrast to earlier discussions and even the formal legislative proposals, the 10 per cent as finally agreed related not to biofuels, but to all renewable energy used in transport (even if biofuels would still dominate). Thus the member states committed to mandatory 20 per cent/10 per cent targets, but ensured they retained considerable flexibility over how those targets were to be achieved. As a final observation on this, more than one interviewee indicated that the 10 per cent figure was agreed before any discussion was held about how it might be achieved.

Another dimension to this debate emerged from our interviews with senior policy insiders in Brussels.[20] By the mid-2000s EU car-makers were coming under pressure because of the failure to deliver on the (voluntary) biofuel blending targets contained in the Biofuels Directive. In response, as plans for the RED took shape, car industry representatives pushed for a wider climate change and emissions-reduction policy, so as to share the burden of meeting the mandates with other sectors and renewable energies. It is also surely no coincidence that the Industry Commissioner, a German, also initially opposed mandates, although interviews suggested his position changed when he realised that failure to act would see EU car manufacturers, over time, fall behind those from other countries.

Figure 3.1 shows countries' progress towards the 2020 targets, both for renewable energy in gross final energy consumption and renewable energy in transport fuel. This shows just how variable different countries' progress has been thus far, with progress on the transport fuel target the more uneven of the two.[21] Interestingly (especially in the

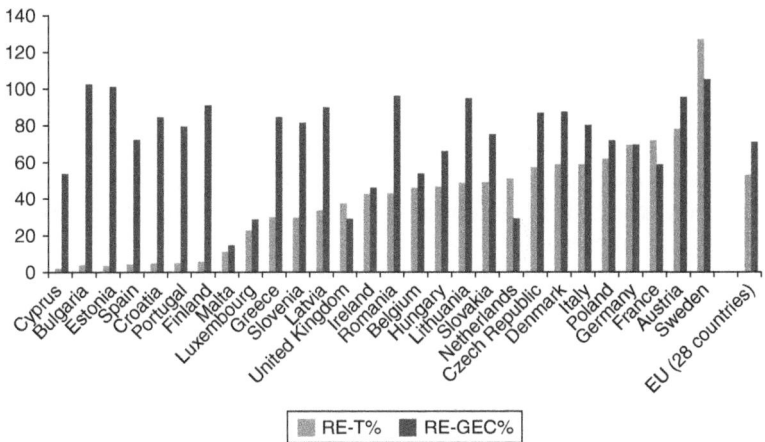

Figure 3.1 EU member states' progress towards their 2020 renewable energy usage targets: 2012, per cent of 2020 targets
Notes:
RE-T: Renewable Energy in Transport Fuel.
RE-GEC: Renewable Energy in Gross Final Energy Consumption.
Source: Eurostat; own calculations.

context of the discussion of US policy in Chapter 4), Sweden has by far the largest FFV car fleet in the EU, as well as a large ethanol powered bus fleet – allowing greater ethanol penetration per vehicle in the transport fleet.

Delivering on the three policy drivers – The challenges of GHG emissions

The previous section charts in some detail the origins and development of biofuels policy in the EU, through to the RED and FQD. All three drivers of biofuels policy have, at various times during this process, been used to justify EU policy. The production of biofuels, and their blending into fossil-based transport fuels, contributes to energy security, by diversifying energy types and source countries. Depending on exactly how one defines rural development (see Chapters 1 and 5), it can be argued that biofuels bring jobs to rural areas and offer a different market for farmers. At various times, several of the Commission documents referred to earlier also made the point that efforts to reduce GHG emissions would require investment in new technologies, which would also generate new jobs and economic activity. The biggest challenge for biofuels policy, therefore, is to deliver on the goal that has been most prominent in EU biofuels policy discourse – the delivery of environmental benefits, in the form of GHG emissions reductions. Here, we introduce the policy instruments used to deliver on the EU policy targets, reflecting also the need to try to avoid the potential downsides of first generation biofuels, outlined in Chapter 1.

EU documents, from 2004 on, make repeated reference to the need to ensure that biofuels deliver lower GHG emissions than the fossil fuels they are replacing. This is based on the recognition, discussed in Chapter 1, that even amongst first generation biofuels, different feedstocks, produced in different ways, on different types of land, and processed using different technology pathways, are capable of delivering widely varying GHG emissions performances. This applies a fortiori when comparing first generation and advanced biofuels, given that the latter avoid food v. fuel conflicts, with many also avoiding land-use concerns. On the other hand, as indicated above, in order to achieve political agreement on the overall targets, a compromise was necessary by avoiding any formal subdivision of those targets by biofuel type (in stark contrast to the US approach, discussed in Chapter 4).

This concern notwithstanding, the EU approach demands ever-greater GHG emissions reductions from biofuels, with an overall cut in GHG emissions by 2020, of 6 per cent, relative to a base of 2010 (Article 7a of the FQD). In addition, the FQD also sets out possible additional non-mandatory emissions reductions, of a further 4 per cent in total, utilising, for example, carbon capture and storage technologies, electric road vehicles and the purchase of credits through the Clean Development Mechanism of the Kyoto Protocol (see Articles 7a and 9). On this point, it is worth noting that the Environment Directorate-General (DG-ENV) was the pen-holder on the revision of the FQD and (interviews indicated) it was they who, from the outset, had to push against the energy interests in DG TREN to get explicit emissions-reduction targets included. In this context, the fact that this 6 per cent reduction target is included in the FQD but not the RED is interesting, given that whilst DG-ENV was pen-holder for the FQD, DG-TREN was pen-holder for the RED.[22]

Whilst this target is laid down in the FQD alone, a number of (more widely reported) targets are present in both the FQD and the RED. These are not compulsory targets for biofuels blended in the EU; they are, however, targets which must be delivered on for the blender to be eligible for fiscal incentives. Only these biofuels can count towards the 10 per cent blending mandate. The reasons for not requiring all biofuels sold in the EU to conform with these standards are analysed in Chapter 8.

As originally agreed, biofuels would have, initially, to deliver at least 35 per cent lower GHG emissions than the fossil fuels they are replacing, a target which applied immediately for production facilities opened in 2008 or later. Older facilities have had, until 2013, to ensure they could deliver this figure. From 2017 this figure rises to 50 per cent with, from 2018, biofuels produced in new facilities having to deliver GHG emissions savings of at least 60 per cent. Despite the different emissions performances of different biofuels, and despite the absence of an explicit target for different types of biofuel to be delivered to market, these emissions targets, indirectly, serve a similar purpose. Specifically, poorly performing biofuels are not eligible for fiscal support at all, whilst the higher target from 2017 would have a dramatic impact on those biofuels with intermediate GHG emissions reductions performance. These targets and schedules are currently up for reform. We consider these reform proposals further below.

It is worth reiterating that whilst these thresholds apply to all biofuels, the higher targets scheduled currently to be introduced in 2017 and 2018 are most likely to be binding on first generation biofuels. This, of itself, is a policy incentive, albeit a weak one, for the development and commercialisation of advanced biofuels. To add to this, the policy laid down that all advanced biofuels would count double in terms of their contribution to the 10 per cent share target for renewables in transport fuels. The absence of any explicit apportioning of the mandate between first generation and advanced biofuels, however, limits the extent to which the private sector is willing, let alone able, to respond to this by sustained investment in the necessary research and development of commercial-scale production.

Sustainability criteria and ILUC: The salami-slicing of difficult issues

EU biofuels policy is, in effect, bifurcated into two: efforts continue to promote and develop advanced biofuels, whilst trying to manage the potential downsides of first generation biofuels. Two interlinked issues have been of particular concern to policy-makers. The first is how to stop the 'wrong' type of land being used to grow biofuel feedstocks. The policy response has been to lay down a series of exclusionary sustainability criteria. They identify land-types which, if used for biofuels sold in the EU, would render those biofuels ineligible for fiscal incentives and would not count against the mandate.[23] The criteria address two distinct issues: biodiversity and emissions, the latter being where the cultivation of the land would, of itself, release carbon stored in the soil. The criteria are shown in Box 3.1:

Box 3.1　EU sustainability criteria (exclusionary)

EU legislation sets out what does not count as sustainable for biofuels production. Biofuels feedstock production cannot occur on certain types of land with a specific function or status before 2008. Lands excluded for biodiversity reasons are:

- primary forests and woods, undisturbed or lacking 'visible' human activity;
- land protected under law, international or intergovernmental agreement (unless feedstock production does not compromise the nature-protection goals);
- highly biodiverse grassland (except, for 'non-natural' grassland, if biofuel feedstock harvesting is required for grassland status to be maintained), although by the end of 2010, the Commission has still to produce a definition of highly biodiverse grassland.

Certain types of land are excluded if carbon would be released if disturbed by feedstock production:

- wetlands;
- continuously forested area;
- undrained peatland (unless feedstock production and harvesting do not require the land to be drained).

This gives rise to a second set of concerns: how far should land-use change considerations be incorporated? The concepts of direct and indirect land-use change were introduced in Chapter 1, with ILUC analysed further in Chapter 9. The question of ILUC presents multiple challenges to policy-makers. Should it be included in emissions calculations for the purposes of the GHG emissions-reduction calculations? If yes, how can we model it, given that it cannot be observed directly? Furthermore, how 'indirect' should we go in these calculations? The EU response to date has been one of the steady development of policy, focusing on tractable problems – even though, as we shall see, the solutions agreed or proposed continue to divide opinion.

Given the policy challenges the EU faced, it did not attempt to negotiate everything at once. Instead, it has pursued what we can call the salami-slicing of policy (Zahariadis, 2003: 15). This is a simple idea which describes how, when faced with a very challenging policy issue, policy-makers progress a step at a time. First, the RED and FQD, enacted in 2009, contained the sustainability criteria. It was over a year later, in June 2010, when detailed rules on the implementation of the sustainability criteria were agreed. ILUC-related proposals should have been presented by the end of 2010, but ongoing debate (and the fail-ure of the Commission to complete the Impact Assessment on time) resulted only in an initial overview of possible policy responses pre-sented by that date. The formal proposals were presented in October 2012 but by December 2013, with no agreement forthcoming, they were shelved. They are expected to be picked up again in late 2014 or early 2015, but with 2014 marking both European Parliament elec-tions and the handover to new Commissioners, even early 2015 may be optimistic.

Even with this stepwise process of policy-making, significant chal-lenges have been faced. Four months before the final implementing rules for the sustainability criteria were published,[24] a version was leaked to Friends of the Earth Europe which, whilst confirming continuously forested areas were to be defined as land with high carbon stock, also stated (on page 9, emphasis in original) that *'a change from forest to oil palm plantation would not per se constitute a breach of the criterion'*. In other words, cutting down virgin rain forest and replacing it with a palm oil plantation might still enable any resulting biodiesel to count as 'sustainable'. This clause was not included in the implementing rules as finally published.

Interestingly, in those final rules the terms 'DLUC' and 'ILUC' are not used. Instead, reference is made in Annex II simply to 'Land Use Change'. Here, the Commission is clear that LUC is determined with reference to the IPCC definitions of land use, as set out in Chapter 1. It is also the case that, for the purposes of implementing the EU sustainability criteria, a change within a category, notably between crops within 'cropland', is not considered LUC as it would not trigger carbon emissions.[25]

The next challenge was how to deal with ILUC which, already by this time, had taken on the status of an issue around which dis-parate anti-biofuels interests could coalesce. The Council and European

Parliament, in both the RED and FQD, called on the Commission to develop a methodology for calculating GHG emissions resulting from ILUC, including the possibility of introducing an ILUC factor into those calculations. With detailed analytical work still ongoing the report that was published (European Commission, 2010: 14) laid out four possible policy options that the Impact Assessment would address:

(1) take no action for the time being, whilst continuing to monitor,
(2) increase the minimum greenhouse gas saving threshold for biofuels,
(3) introduce additional sustainability requirements on certain categories of biofuels,
(4) attribute a quantity of greenhouse gas emissions to biofuels reflecting the estimated indirect land-use impact.

In the public consultation process for this report, two countries' responses are worthy of note. Indonesia and Malaysia (major producers of palm oil and the focus of much attention over deforestation) both called for no action currently, picking up on the point that the increased GHG emissions thresholds from 2017 and 2018 would address any potential ILUC impact. Malaysia also referred to continued problems with the accurate estimation of ILUC-related emissions (an issue analysed in detail in Chapter 9).

As a footnote to this section, we consider briefly the troubled history of ILUC in EU policy development, as revealed through our interviews. Notably, prior to the publication of Searchinger *et al.* (2008), Tim Searchinger gave a seminar in Brussels, which brought the potential magnitude of ILUC as an issue to the attention of officials, even before the research was published. Despite this, the proposals continued to be developed without recognition of ILUC as a factor or concern. This also, one can argue, helps explain the delay in getting ILUC matters addressed formally by the Commission. It was, we were told, only once the proposals got into co-decision that some in the European Parliament, and some member states, picked up on the need for ILUC to be recognised. This, it was argued, led to the Commission committing to report on ILUC by 2010. It also led to an inclusion of a bonus in the emissions calculations, of 29 gCO_{2eq}/MJ for biofuels grown on degraded or idle land (see Chapter 1). This, one interviewee argued, was purely a political fix, aimed at helping to address the ILUC question sufficiently to get the RED and the Climate and Energy

Package through on schedule – there was no scientific basis for its inclusion.

Proposals for reform of the RED and FQD

Recognising a number of these pressures, legislative proposals for reform of the RED and FQD were published in October 2012 (European Commission, 2012). They include a proposal for dealing with ILUC, based on the application of an ILUC factor: 12 grams of CO_2-equivalent per megajoule of biofuel energy for cereals and other starch-rich crops, 13 for sugars, and 55 for oil crops. In addition, changes have been proposed to the dates when the higher GHG emissions thresholds will apply.

New bio-refineries, in the published reform proposals, would have to deliver 60 per cent GHG emissions reductions from 1 July 2014, compared with 1 January 2017 currently – although the failure to agree the reform by the end of 2013 (at which point the proposals were set to one side for at least one year) means the 2014 date would have to be moved back. Moreover, refineries in operation on or before 1 July 2014 will have a full extra year, until 1 January 2018, to deliver a 50 per cent GHG emissions reduction, although this would now have to include the new ILUC factor.

A third element is the proposal to place a cap on the share of the renewable fuel mandate that can come from first generation biofuels. The proposal is 5 per cent, roughly equal to the actual share of first generation biofuels in total transport fuel in 2012. The 10 per cent overall target remains, but given the superior GHG emissions performance of advanced biofuels, the multiplier applied to some types of advanced biofuels in calculating their volumetric contribution towards the 10% target figure is to be increased from a factor of 2 to 4. Thus there are further efforts to encourage the development and commercial production of advanced biofuels, with part of the biofuels mandate now given over explicitly to them. That said, in conversations with leading figures from fuel companies, we were told that these proposals would have no impact on encouraging the development of advanced biofuels, because they provided no guarantees regarding returns on any investments made. Indeed policy instability and change, of itself, is having a highly damaging impact on investment by the EU biofuels industry.[26]

Negotiations on these proposals ended in December 2013, with an expectation that discussion would not resume until late 2014 or 2015.

Thus the (possible) introduction of an ILUC factor has been delayed, and the changes to GHG emissions threshold minima will, with the delay in agreeing reforms, be at most only marginally different from the existing policy. The state of negotiations in 2013 suggested a compromise figure of a 7 per cent share of first generation biofuels in the total renewable target was a plausible outcome – which is unlikely to be significantly below what is possible by 2020 anyway. Thus, of the three elements of the reform proposal discussed, it is only the possible introduction of the ILUC factor that would appear to represent a notable shift, in terms of impact, from the current policy – and even here the impact is likely to be attenuated by the already-scheduled rise in GHG emissions thresholds.

Looking back on the evolution of EU biofuels policy, it emerged from the EU's positioning of itself as a global climate and environmental policy leader. As noted earlier, however, research such as that by Skovgaard (2013) suggests, more recently, leadership role has been diminished, notably through the greater engagement of economic interests. Afionis and Stringer (2012) look at this broad question specifically in the context of biofuels policy. They find that in this specific setting, a similar pattern has emerged. They interpret trade barriers erected to protect inefficient EU biofuels producers as indicative of a policy where economic interests have been placed above environmental goals, attenuating the delivery of the latter. It remains to be seen, however, if the influence of economic interests over environmental policy goals is a short-term consequence of the prevailing economic conditions during and immediately after the global economic crisis, or whether it represents a longer-term loss of the conditions for policy entrapment.

Conclusions

EU policy emerged relatively late on, compared with even the US (as discussed in the next chapter), let alone Brazil. Some EU member states had active policies in the distant past but, for a variety of reasons, these lapsed. The current EU biofuels policy, dating from the early 2000s, has clear antecedents in wider environmental and climate change policy developments which began in the 1980s. Even though energy security and rural development drivers are both present in EU policy, it is the climate change mitigation driver of biofuels policy that has been pre-eminent.

This has, however, only served to magnify the attention placed on the emissions-related aspects of EU policy, in particular the external consequences of the policy. We look at these in more detail in Chapter 8.

At this stage, we can highlight that whilst the policy has a number of elements that seek to promote the development, production and use of advanced biofuels, most policy attention has been directed towards putting in place measures which contain the potential downsides of the first generation biofuels, currently the dominant product of EU biofuels policy.

4
US Biofuels Policy – An Introduction and Overview

Whilst all three drivers of biofuels policy identified in the previous chapters are present at different times and in different combinations in the US case, the particular influence of energy security considerations makes US biofuels policy distinctive. Of the three drivers, energy security has been the most enduring in the discourse and outcomes of US biofuels policy. The origins of current policy can be traced to the oil price shocks of the 1970s that revealed the macroeconomic vulnerability of the US resulting from its dependence on imported oil. Whilst sensitivity to the economic and national security consequences of importing significant shares of US total energy consumption diminished somewhat in policy-making calculations in the following decades, the 9/11 terrorist attacks in 2001 and – in particular – the Iraq war from 2003 set a context for energy security concerns to be a central influence in the significant US biofuel policy reforms of the last decade. These reforms established the backbone of goals, instruments and settings in current US biofuels policy.

Before proceeding, clarification of an important limitation in scope of the chapter with respect to US federalism is required. The chapter presents an overview of the development of US biofuels policy at the federal government level; it does not consider the patterns of state-level policy instruments and settings that operate to affect the incentives – with varying success – in the economics of US biofuel consumption and production. For example, as of 2013, California has 27 laws and regulations in the area of alternative fuels and vehicles, alongside 29 incentive programmes offering a variety of grants and tax credits for their use. The US Department of Energy (2013) has maintained a database of both federal and state incentives for alternative fuels, including tax reductions and exemptions which indicate the scale and variety of subnational biofuels policy that is not covered here. In particular, we do not consider the actual market consequences of the interaction between the US federal-level policy framework and state-based initiatives.

Nevertheless, we believe this narrowing of scope is justified in terms of both the practical constraint of space as well as, importantly, the overarching ambition of the chapter: to articulate how the three drivers of biofuels policy processes have played out in different and highly specific ways in the US context and understand how, why and when reform has occurred. We do, in Chapter 8, look at how Californian State and US Federal policies have interacted to create a particular and peculiar pattern of trade, but the primary focus in that chapter is on how policy variations at the state and national levels have combined to produce a peculiar pattern of trade flows for ethanol at the international level.

This chapter first sets out the broader political context against which the different drivers of biofuels policy in the US played out, before going on to detail a chronology of policy development. This longitudinal account is, in turn, used to establish some of the central tensions in the ongoing changes in policy instruments and their settings, which characterise the implementation challenge in current US biofuels policy.

In the shadow of Iraq: The political context for the rush to biofuels

During his second term, President George W. Bush elevated the ambition to increase the production and consumption of biofuels to a central plank of his energy policy. He used his State of the Union addresses, in particular in 2006 and 2007, to set this agenda: profiling his commitment to biofuels, and ethanol in particular, in terms of reducing dependence on imported oil. In the 2006 speech he called for cellulosic ethanol to be cost competitive and on the market with corn-based ethanol by the year 2012, whilst a year later a 20 per cent reduction in absolute terms in US gasoline consumption within 10 years was the headline policy ambition.

This deployment of the agenda-setting power of the US Presidency is an important episode in the unfolding of biofuels policy reforms in the US over the last decade. Although it is Congress that enacted the key pieces of legislation which substantially reformed US biofuels policy in the mid-2000s, the Bush agenda for biofuels was the influential backdrop to how the three biofuels policy drivers have played out in the unique US experience. In particular policy advocacy from the Presidency, in this case calling attention to energy security concerns, is an important strategic resource in any US policy-making process. Its deployment in the biofuels policy domain was significant for the development of US biofuels policy.

Understanding context is important because although the same driver or combination of drivers may seem to operate in the same way in different contexts, it is the attributes of those contexts that will determine the type of policy outcome observed: what is reformed, when and how. It is important to recognise that any account of biofuels change relies on the interaction between the different reforms drivers and the political contexts within which they have operated. Most notably, the energy security driver is sensitive to the broader geopolitical environment and changing ideas about the US national interest.

Much of the political context to recent US biofuels policy is, of course, well-known. Against a backdrop of international protests, in March 2003 the US led an invasion of Iraq, which although linked to the 11 September 2001 attacks by some, was more loudly claimed to be holding stockpiles of weapons of mass destruction. Military-wise, the initial invasion had seemed successful and within two months President Bush announced the end of major combat operations in Iraq. However, in the following months, insurgents began resisting US occupation. Additionally, historic religious tensions between majority Shiite and minority Sunni Muslims manifested as sporadic but intense violence. By the end of 2003, despite the war being initially popular in the US, the post-war occupation had begun to lose public support in the US and as a corollary President Bush's job approval rating began to fall.

Despite this, in 2004 Bush narrowly won re-election against Senator John in a campaign dominated by national security issues and the war in Iraq rather than domestic issues. Bush began his second term against a backdrop of increasing violence in Iraq that threatened US control of Baghdad as well as full scale, country-wide civil war. The domestic unpopularity of US involvement in Iraq increased markedly as it became clear that the country had held no weapons of mass destruction at the time of invasion. Both President Bush's job approval rating and the popularity of the Republican Party, more broadly, declined steeply at the outset of his second term. The Democrats gained majority control of both the House and the Senate in the 2006 mid-term elections, running a successful campaign against a 'do-nothing' Republican-controlled Congress as well as a 'lame duck' Presidency.

In the specific politics of biofuels policy, two landmark pieces of legislation were passed against the backdrop of change in the US political context. The 2005 *Energy Policy Act* introduced a Renewable Fuels

Standards (RFS) programme to encourage the use of ethanol and biodiesel with the goal of doubling their use by 2012.[1] In 2007, the *Energy Independence and Security Act* (EISA) introduced RFS2, which requires the incorporation of 36 billion gallons of ethanol by 2022.[2] EISA also limited the amount of corn-based or conventional ethanol to 15 billion gallons of that total. The remainder is expected to be made up from growth in cellulosic ethanol, biodiesel and other unspecified advanced biofuels, as shown in Table 4.1. We discuss these mandates further below.

As for where the 36 billion gallon figure came from in the first place, interviews indicate the starting point was, as noted earlier, the goals for biofuels laid out by President George W. Bush in his State of the Union addresses. This, it was estimated, would translate to about 35 billion gallons – and so, with Senate controlled by the Democrats (and with the Senate pushing for a revised RFS to be included in what became the EISA), a figure was arrived at that surpassed that suggested by the president. That said, another interviewee suggested that Congress did

Table 4.1 The US Renewable Fuel Standard (billion gallons)

Year	Conventional	Total advanced biofuel	o/w Cellulosic	o/w Biomass-based diesel	o/w Non-cellulosic advanced	Total
2008	9					9
2009	10.5	0.6		0.5	0.1	11.1
2010	12	0.95	0.1	0.65	0.2	12.95
2011	12.6	1.35	0.25	0.8	0.3	13.95
2012	13.2	2	0.5	1*	0.5	15.2
2013	13.8	2.75	1	1	0.75	16.55
2014	14.5	3.75	1.75	1	1	18.15
2015	15	5.5	3	1	1.5	20.5
2016	15	7.25	4.25	1	2	22.25
2017	15	9	5.5	1	2.5	24
2018	15	11	7	1	3	26
2019	15	13	8.5	1	3.5	28
2020	15	15	10.5	1	3.5	30
2021	15	18	13.5	1	3.5	33
2022	15	21	16	1	4	36

Note: *The Environmental Protection Agency shall, each year from 2012, set a figure of at least 1 billion gallons.
Source: Based on data in US Senate and House of Representatives (2007).

not realise the problems, discussed below and throughout this book, that would result from the 36 billion gallon figure.

A chronology of US biofuels policy

Contemporary US biofuels policy can be traced back to the introduction of various incentives for ethanol production in the second half of the 1970s. Against the backdrop of the adverse macroeconomic shocks in the US induced by spikes in international oil prices, the *Energy Policy Act* 1978 introduced a tax credit for blenders of transport fuels as a policy instrument, initially equivalent to a subsidy of 40 cents per gallon of ethanol. A similarly structured tax credit in the much smaller biodiesel blending market was not put in place until the mid-2000s. The level of tax credit for ethanol blending varied over 40 years before being abolished at the end of 2011. Support for the ethanol industry has been the subject of much public politics in the US – indeed the famous, four-yearly ethanol pledge in the Iowa Presidential Primary was the subject of an episode of award-winning TV series, *The West Wing*.

To complement the introduction of the tax credit, a tariff on imported ethanol was also introduced in the late 1970s. The regular tariff of 2.5 per cent was supplemented by a category known as 'other duties and charges' in order to offset the effect of the tax credit to blenders. This was justified on the grounds that the ethanol tax credit to blenders could be applied to both domestic and imported ethanol, and US policy should not subsidise the use of imported ethanol. Biodiesel was also protected by a tariff on imports into the US of 4.5 per cent ad valorem from the early 2000s.

Throughout the history of the ethanol tax credit and tariff, opportunities have existed for preferential access to the US market. *The Caribbean Basin Economic Recovery Act* (CBERA) allows duty-free import of ethanol to the US if at least half of it uses feedstock from any of the CBERA partner countries. Brazil has been able to supply the US market despite the tariff; for example, ethanol can enter the US duty free each year under Central American Free Trade Agreement (CAFTA), even if the feedstock for which is originally sourced from Brazil.

Several interviewees for this book linked the development of the US ethanol industry to the introduction of the tax credit and tariff policy instruments from the 1970s. Corn became the predominant feedstock for ethanol production because of its relative abundance in the US as well as the existing, well-known technology for converting it into alcohol. Policy introduced in the 1970s helped ethanol production to remain viable even as ethanol prices softened along with crude oil and gasoline prices through the 1980s.

Although energy security imperatives were the catalyst to the biofuels policy reforms introduced in the 1970s, biofuels policy has been and still is conventionally regarded as part of established agricultural policy (interviewees in the US consistently referred to the role of 'Big Ag' in the origins of ethanol policy). For example, biofuels policy in the late 1970s was linked explicitly to the development of the so-called 'Minnesota Model' for ethanol production, in which farmers were encouraged to produce ethanol to add value to their corn (Bevill, 2008). The Minnesota Model was a template for achieving agreement between interested parties in biofuels to collaborate, in order to retain profits in rural communities through employment opportunities and adding value to existing agricultural products. In addition to the introduction of the policy instruments already noted, the development of the ethanol industry was also indirectly stimulated by the high support price of sugar that led to the growth of corn milling capacity to produce high fructose corn syrup: the same plants are capable of producing ethanol.

Biofuels policy settings remained relatively unchanged during the 1980s and 1990s. However, the period was notable for the increasing use of ethanol as an oxygenate blend in conventional transport fuels to control carbon monoxide emissions. Amendments to the *US Clean Air Act* 1990 mandated oxygenates be used in fuel to improve air quality, and ethanol was prescribed as one option along with Methyl Tertiary Butyl Ether (MTBE).

An unintended consequence of the Clean Air Act amendments was the emergence of concerns about groundwater and soil pollution caused by MTBE. These became increasingly loud from the late 1990s on, with about half of the states in the US subsequently banning it or restricting its use. The *Energy Policy Act* of 2005 did not ban the use of MTBE, but the Environmental Protection Agency (EPA) removed MTBE liability protection. Thus the policy change on the use of MTBE triggered by the EPA decisions became institutionalised. Interviews for the book revealed that it was lobbying from the corn and ethanol sectors that protected the phasing out of MTBE as a fuel blend from being reversed by subsequent policy interventions.

More generally, a coalition of Big Ag interests developed, concerned with the increased use of renewable transport fuels, given that the 2005 changes regarding MTBE also saw the removal of the oxygenate requirement, which represented a lost market for ethanol. They sought to exploit the window of opportunity provided by a policy context favouring proposals to reduce dependence on imported oil, to push for policy instruments that would not just maintain, but increase significantly the demand for ethanol. In the *Energy Policy Act* 2005, the first Renewable Fuels Standard (RFS) was introduced. It provided for ethanol production of four billion gallons in 2006 with an increase to seven and one-half billion gallons by 2012. On the back of the Republican defeat in 2006, one of the first pieces of legislation passed by the Congress was the *Energy Independence and Security Act* 2007 that required renewable fuel usage to increase to 36 billion gallons annually by 2022. As of early 2014, this second RFS (RFS 2 hereafter) still guides national ethanol policy, although the figures as shown in Table 4.1 can be amended each year by the EPA, in line with actual supply potential.

Although US policy reform in this period referenced climate change mitigation, for example, in the 2007 State of the Union addresses of President Bush, it was far less prominent in public as well as insider deliberations than the other drivers. One interviewee in Washington went so far as to say that only after 2008 could climate change be spoken of publicly as important for US policy. As noted, in a conventional static analysis, biofuels policy belongs as part of agricultural policy domain. This is seen in RFS2 with the de facto ring-fencing of 15 billion gallons of ethanol, principally, for corn. However, when the dynamics of US biofuels policy are considered, it is the energy security driver that furnishes the ability of biofuels questions to achieve a wider prominence, and grab attention in the core executive of the federal government ahead of other policy domains.

End of the ethanol tax credit

For much of their history, the ethanol tax credit and tariff established in the 1970s had appeared to be inviolable. Indeed, interview data collected in 2010 and 2011 revealed a widely held assumption that it was an enduring constraint on future US biofuels policy. The ethanol tax credit had been reduced from an equivalent rate of subsidy of 54 cents per gallon in the 15 years until 2005, when it was reduced to 51 cents. A lower rate of 45 cents came into effect in 2009 but there was no widespread expectation that these reductions augured the removal of the tax credit entirely.

However in 2011, when the Congress, and broader US political system, was preoccupied with the looming fiscal cliff of expiring tax cuts and planned public expenditure cuts, both the tax credit and tariff instruments in biofuels policy were removed. An unlikely and short-lived coalition of fiscal conservatives and environmental groups emerged to frame them as emblematic of corporate welfare and the reasons for the structural deficit in the US federal budget. The Congressional Budget Office (CBO) had put the annual cost of the tax credit in 2009 at $6bn (CBO, 2010), whilst a 2010 study by the Environmental Working Group of Congress (2011) estimated that the cumulative fiscal cost of ethanol subsidies between 2005 and 2009 was $17 billion, likely to rise to US$53.59 billion if the tax credits were extended until 2015. The 2010 CBO study had also estimated that taxpayer costs to reduce gasoline consumption by one gallon were $1.78 for corn ethanol and $3.00 for cellulosic ethanol.

In a similar way, and without considering potential indirect land-use effects, the costs to taxpayers of reducing greenhouse gas emissions through tax credits were about $750 per metric ton of CO_2-equivalent for ethanol and around $275 per metric ton for cellulosic ethanol. In an influential assessment of the RFS published by the National Research Council (2011), on the then current policy settings, net budgetary costs to the federal government were likely to balloon mostly as a result of increased spending on payments, grants, loans and loan guarantees to support the development of cellulosic biofuels and foregone revenue, as a result of biofuel tax credits.

In June 2011, the US House of Representatives approved an amendment to an economic development bill to repeal both the tax credit and the tariff. This bill ultimately did not move forward into law, but signalled publicly for the first time the shifting coalitions in US biofuels policy. By the end of the year Congress, through a non-decision, removed these two long-standing policy instruments by not agreeing to extend either beyond their already legislated expiry of 31 December 2011.

A deteriorating fiscal situation was the proximate catalyst for this historic defeat of the ethanol coalition in 2011. The *Budget Control Act* of 2011 has been signed into law by President Obama on 2 August 2011. The Act brought conclusion to the 2011 US debt-ceiling crisis, whilst similar fiscal events followed in 2012 and 2013, which had threatened to force the US government to default on its debt repayment obligations. The law introduced several public expenditure management mechanisms, such as creation of the Congressional Joint Select Committee on Deficit Reduction, that serve to draw attention to – amongst other things – large, expensive parts of industry policy.

The expiration of the tax credit and tariff also reflected the shifting coalitional dynamics, as an effect of relatively high corn prices as well as relatively strong markets for conventional corn ethanol. Within the pro-biofuels coalition there was broad, if reluctant, acceptance of the proposition that with a binding RFS in place and a slow-moving but fully anticipated movement towards the blendwall constraint, the tax credit and tariff were to a large extent redundant, or at least not central to their lobbying interests. The operation and future of the RFS, and the blendwall issue, is considered below.

Although this chapter does not cover the emergence of hydraulic fracturing or 'fracking', this is also relevant in tracking the influence of the energy security imperative in the recent US biofuels policy change. The contextual influence of the estimates that US will be self-sufficient in energy in the next 20 years (see, for example, BP Energy Outlook, 2014) is difficult to pin down precisely, and is undoubtedly a second-order effect in much of the 2011/12 biofuels politics. Since its widespread adoption in the last five years, hydraulic fracturing technology – which allows drillers to extract gas and oil from shale deposits using water pressure – has reshaped the US energy landscape. According to the US Department of Energy, over two million oil and gas wells have been hydraulically fractured, and that currently about 95 per cent of completed oil and gas wells in the US are being hydraulically fractured (US Department of Energy, 2013).

Hydraulically fractured wells now constitute well over 40 per cent of domestic oil production, as well as two-thirds of the current natural gas production in the US. Domestic oil production has increased by 43 per cent, to 7.3 million barrels a day, since the introduction of EISA in 2007. Combined with increasing fuel efficiency, net imports of crude oil have fallen rapidly from 11 million barrels a day to 6.5 million between 2006 and 2011, a trajectory that is expected to continue. Exports from the countries of the Organisation of Petroleum Exporting Countries (OPEC) to the US fell by a third in the five years to 2012, whilst US imports from Persian Gulf countries are down to 2.1 million barrels a day. The additional 2.2 million barrels a day added to global supply since 2008 though boosts in US production seems to have decoupled, or at least lessened the strength of, the link between political crises in the oil exporting parts of the Arab world and price volatility on world markets. For example, oil production in Libya dropped vertiginously in 2011 without causing the expected spike in oil prices.[3]

Whilst it is tempting to cast recent adjustments in US biofuels policy instruments in 2011 and a shift in policy discourse in the light of a reshaped energy policy landscape, the causal chain between the two is extended and opaque in parts. Nevertheless, as a general proposition the prospect of energy self-sufficiency depowers the energy security driver, and though the energy security driver is not removed from US biofuels policy-making, its influence is attenuated. This proposition holds a fortiori in a broader fiscal policy situation, in which the risk of sovereign default is being mentioned publicly, with all its possible attendant consequences for the world financial system.

Whilst the future of US energy supply remains subject to some risk and uncertainty, it still seems that biofuels will play a key role in the energy mix. Overall, biofuels contribute over 20 per cent of all renewable energy consumed in the US. Other than hydro-electric power, it is the largest single contributor, the same as wood and slightly more than wind (US Department of Energy, 2013). Further, it is not clear how a putative diminution of the energy security driver affects the interpretation of the other drivers for a continuation of current US biofuels policy.

The biodiesel tax credit still standing

The story of the biodiesel tax credit, which operates in the same way as the ethanol tax credit, is slightly different. US biodiesel production and consumption remains relatively modest compared to the EU, and also compared to US ethanol production. However, it has grown significantly in the last decade and presents one of the clearer successes of the advanced biofuels part of RFS2. Initially established by the *American Jobs Creation Act* 2004, the biodiesel tax credit was subsequently extended by *the Energy Policy Act* of 2005 as part of the original RFS. Although it was allowed to lapse briefly in 2010, it was subsequently retroactively extended until the end of 2013. The tax credit is currently (end of 2013) $1.00 per gallon biodiesel tax credit for producers or blenders of pure biodiesel, biomass-based diesel or diesel/renewable diesel blends.

Like the ethanol tax credit in the 1970s, the biodiesel version has been a critical factor in the growth of the biofuels industry. From a very low base, biodiesel production has grown rapidly in the last decade, a trend that has been boosted further recently as part of market adjustments to the blendwall policy dilemma discussed below. In addition, there was a period of exporting from the US to the much larger EU biodiesel market that sparked a trade dispute. US exports were encouraged by the fact that biodiesel producers received a US tax credit even on exported fuel, and received tax credits in the EU as the product qualifies as biofuel. The European Commission imposed anti-dumping and countervailing duties on such imports in March 2009 and in July 2009 announced a temporary tariff to be in effect for five years.

Whither the RFS?

Both the 2005 *Energy Policy Act* and 2007 EISA delegated the powers of establishing and implementing regulations relating to RFS1 and RFS2 to the EPA. The EPA is responsible for ensuring that the US transportation fuel supply contains the mandated biofuel volumes. These mandates are set annually by the EPA as the minimum volume of biofuels is to be used in the national transportation fuel supply each year on a rising trajectory towards the legislated 2022 volumetric targets. The EPA's initial regulations for implementing RFS1 established detailed compliance standards for the whole biofuels supply chain. The formal legal obligation is imposed on gasoline producers (mainly the refiners and importers), in the form of a renewable volume obligation (RVO). In order to meet this requirement, which increases year by year as set by the EPA, producers need to buy renewable identification numbers (RINs) generated by blenders of biofuels and gasoline. This compliance with RFS volume obligations is part of a producer's licence to continue to operate as a gasoline refiner or importer.

RINs are the heart of the tracking system for the renewable fuel portion of the US transport fuel. The price of RINs will be reflected in the price of gasoline available as blendstock in wholesale markets, since it is part of the regulatory cost for gasoline producers. For blenders operating in wholesale markets, generation of RINs will be a core requirement for them to be able to secure the regular supply of gasoline from refiners and importers.

The EPA rules for administering RFS2 built upon the earlier RFS1 regulations, but with several important differences. As already noted, the mandated volumes are significantly higher in RFS2 than RFS1 and the timeframe over which the volumes increase is extended through to at least 2022. The total renewable fuel requirement is divided into four separate, but nested categories – total renewable fuels, advanced biofuels, biomass-based diesel and cellulosic ethanol – each with its own volume requirement (see below). Importantly in terms of policy implementation, biofuels qualifying under each category must achieve certain minimum thresholds of life-cycle GHG emission reductions, with certain exceptions applicable to existing facilities. Finally, under RFS2 all renewable fuel must be made from feedstock that meet a new definition of renewable biomass, which includes not just the types of feedstock that can be used, but also the land that these feedstocks are grown on. Thus, in contrast to the exclusionary EU sustainability criteria discussed in Chapter 3, the US criteria are inclusionary:

Box 4.1   US sustainability criteria (inclusionary)

- Planted crops and crop residue harvested from agricultural land cleared or cultivated before 19 December 2007 either actively managed or fallow, and non-forested;
- Planted trees and tree residue from actively managed tree plantations on non-federal land cleared before 19 December 2007;
- Animal waste material and animal by-products.
- Slash and pre-commercial thinnings that are from non-federal forestlands, excluding forests or forestlands that are critically imperiled, imperiled or rare; and old growth or late successional forest;
- Biomass obtained from the immediate vicinity of buildings and other areas regularly occupied by people, or of public infrastructure, at risk from wildfire.
- Algae.
- Separated yard waste or food waste, including recycled cooking and trap grease.

Table 4.1 set out the numbers in RFS2. Here we elaborate further on the categories of biofuel included. Each category has a specific volume mandate and life-cycle GHG emission reduction threshold. The following four categories are also subject to the new renewable biomass feedstock criteria in Box 4.1.

A. *Total Renewable Fuels*: The total volumetric amount of biofuels mandated, which grows from nearly 13 billion US gallons (49,000,000 m³) in 2010 to 36 billion US gallons (140,000,000 m³) in 2022. Most biofuels, including corn-starch ethanol, fall into this category and must meet a life-cycle GHG emission reduction of 20 per cent. Provisions state that the volume of corn-starch ethanol included under the RFS cannot exceed 12 billion US gallons (45,000,000 m³) in 2010. The cap grows to 15 billion US gallons (57,000,000 m³) by 2015 and is fixed thereafter.

B. *Advanced Biofuels*: Part of the total renewable fuels, this category includes biofuels produced from non-corn feedstocks. This can include sources such as grains like sorghum and wheat, imported Brazilian sugarcane ethanol, as well as biomass-based biodiesel and biofuels from cellulosic materials. Advanced biofuels must reduce life-cycle

GHG emissions by 50 per cent. The mandate grows from nearly 1 billion US gallons (3,800,000 m³) in 2010 to 21 billion US gallons (79,000,000 m³) in 2022.

C. *Cellulosic and Agricultural Waste-Based Biofuel*: Part of the total renewable fuels, this category rises from 100 million US gallons (380,000 m³) in 2010 to 16 billion US gallons (61,000,000 m³) in 2022. Cellulosic biofuels must reduce life-cycle GHG emissions by at least 60 per cent to qualify. Cellulosic biofuels are renewable fuels derived from cellulose, hemicellulose, or lignin. This includes cellulosic biomass ethanol as well as any biomass-to-liquid fuel such as cellulosic gasoline or diesel.

D. *Biomass-Based Biodiesel*. Part of the total renewable fuels, this category grew from 0.5 billion US gallons (1,900,000 m³) in 2009 to 1 billion US gallons (3,800,000 m³) in 2012. Qualifying biofuels include any diesel fuel made from biomass feedstocks including biodiesel (mono-alkyl esters) and non-ester renewable diesel (cellulosic diesel). The life-cycle GHG emissions reduction threshold is 50 per cent.

The RFS is likely to remain central in the future development of US biofuels policy but there remains, as at early 2014, considerable levels of both market and political uncertainty associated with possible spillover effects in other markets (for example, for major feedstocks) and on other important policy goals. Of these uncertainties, the resolution of the policy tension created by the blendwall is paramount. As the *Financial Times* (2013a) put it, 'in the slow-motion collision between US biofuels policy and the "blend wall", it looks like the wall will be left standing'. The blendwall is the label given to the combination of market, legal and technological constraints on the actual demand for US biofuels. It is generally considered in terms of conventional, corn-based ethanol and expressed as an annual figure. The blendwall is a function of lower-than-expected retail demand for petrol, as a result of the shift to more fuel-efficient cars in the US, a secular decline in vehicle miles travelled, as well as the effects of the post-2008 Great Recession, alongside a reluctance for blending companies to increase the ratio of ethanol in each gallon of petrol beyond 10 per cent, citing potential damage to engines.

As of end-2013, the US blendwall is estimated by most market observers at about 13 billion gallons. However RFS2 requires 13.8 billion gallons of conventional ethanol to be blended in 2013, even excluding the advanced categories mandates. Thus a tension exists between what the policy prescribes and the level at which the market, legal and technological factors constrain demand: the blendwall. The continued operation of RFS2 has been questioned; doubts raised about its long-run viability as demand and supply requirements are not in equilibrium; and concerns established in policy discourse about the possibility of unintended consequences of the RFS in other markets.

The greatest unknown for the implementation of RFS2 is the ability of the biofuels industry in the US to meet the expanding mandate for biofuels from non-corn or advanced sources such as cellulosic biomass materials, whose production capacity has been slow to develop, or biomass-based biodiesel, which although it has grown remains relatively expensive to produce owing to the relatively high prices of its feedstocks. Finally, considerable uncertainty remains regarding the development of the infrastructure capacity needed to deliver the expanding biofuels mandate to consumers. These apply in particular to E15 and E85, discussed further below.

The setting of annual mandates by the EPA has been controversial and the source of continued biofuels politics. Whilst the US has moved rapidly towards its corn ethanol target, current progress towards the advanced biofuel targets does not inspire confidence that the 2022 target will be achieved. Critics of the EPA, both in Congress and beyond, have called for the use of the previous year's actual biofuels production totals to dictate new annual targets, rather than backward extrapolation from the 2022 target. A Washington court agreed in 2012 with the American Petroleum Institute (API), a prominent oil industry lobby group, finding that the EPA's mandates were 'unreasonable' (Financial Times, 2013b). The EPA nevertheless increased its mandate for cellulosic ethanol in fuel to 14 million gallons in 2013, a marked increase from the 2012 mandate, of 8.65 million gallons. This is against the backdrop of a disappointingly slow development in the infant cellulosic ethanol industry, much slower than had been hoped for when the RFS2 standard was introduced in 2007 (as we explore further in Part II of the book).

There is now a diverse group of organisations forming a loose coalition to try to have the RFS2 standard lowered, if not scrapped. This follows on from the defeat of the ethanol coalition in 2011 discussed above. The White House in November 2012 turned down publicly a request to repeal the mandate immediately. API President and CEO Jack Gerard responded to EPA's 2013 biofuel mandates under RFS, by promising a renewed campaign for abolition. Whilst the EPA has since 2007 conventionally rebuffed such concerns, in the second half of 2013 its public language shifted to acknowledge the policy dilemma presented by the blendwall. For 2014, an EPA spokesperson was quoted in the *Financial*

Times (2013a) as saying that mandates will be set according to the 'the ability of the market to consume ethanol in higher blends... is highly constrained as a result of infrastructure- and market-related factors'. This was part of an announcement of plans to reduce renewable fuel volume requirements in its rules for 2015.

Key challenges in achieving RFS2 are manifold. First is the high cost of producing cellulosic biofuels compared to petroleum-based fuels and uncertainties in future biofuel markets. Next, as noted, US demand for ethanol – whatever its feedstock – is restricted by the blendwall. The policy response so far has been to seek agreement on raising the ethanol blend in petrol, to 15 per cent, for newer vehicles. Interviews indicated there are several million flex-fuel vehicles in the US, able to run on any ethanol–petrol mix, but mainly because of tax-breaks lowering purchase prices; availability of E85 remains limited and localised.

Given the supply-side constraints discussed above, however, US policy is reaching a critical impasse. The blendwall, combined with improving fuel efficiency and high fuel prices, is limiting demand for ethanol; yet the limited commercial development of advanced biofuels is limiting potential supplies. Interviews with both US agencies and biofuels interests confirmed that the 15 billion gallons devoted to conventional ethanol in RFS2 was the maximum it was believed could be produced without significant impacts on agricultural markets and prices, rather than determined in relation to estimated current demand capacity. It was also determined only to be part of the initial hoped-for supply through to 2022. It is, moreover, also a coincidence that the 15 billion gallon figure is broadly similar to the current blendwall. Thus, in the presence of both demand and supply constraints, the US ethanol market, currently, is approaching a somewhat wobbly equilibrium. The policy, currently, is not delivering advanced biofuels to market, but nor is it able to rectify the demand-side constraint of the blendwall.

Conclusions: Key dilemmas and tensions in policy instruments – Self-undermining dynamics?

Whilst it is tempting to see the shifting energy security picture as diminishing the need for biofuels policy, as part of the explanation of why the ethanol tax credit and tariff were permitted to expire, this is an unduly rational assessment of the US agricultural policy-making process in general, or that for biofuels in particular. The chronology presented in this chapter encourages a more historically orientated analysis, and this final section attempts to shed light on the feedback mechanisms through which biofuels policies seem to have become self-undermining over time in the US, contributing to an increased likelihood of a major change in policy orientation that perhaps the biofuels policy changes of 2011 and 2012 have already signalled.

We try to apprehend a set of endogenous forces – processes deriving from policy itself – that have generated the pressures for biofuels policy change, and expanded the political opportunities for the anti-biofuels coalition. In particular, what is it about the policy instruments detailed above that has created sources of opposition post-enactment, leading to several processes of challenge and gradual pressure to change over time? This self-undermining feedback effect seems to have expanded the coalition of actors willing and able to support biofuels policy reform.

Although biofuels policy implementation is covered in some detail in Part II of the book, it is worthwhile to note in the US chapter that theory offers several conjectures on why losses from the implementation of a particular policy design tend to increase over time (Jacobs, 2011). Two are pertinent to the biofuels case. First, as the layered complexity of a policy regime increases, so do unanticipated consequences beyond the time horizons of the original designers, and these have the potential to be larger losses than expected at the time of the policy enactment. These potential unanticipated losses may serve to recalibrate actors' interests, their policy strategies and overall coalitional dynamics. This is particular acute in the biofuels case where technological and market uncertainty were anticipated and designed into policy after 2007. Secondly, and relatedly, public policies will sometimes present adverse outcomes for particularly powerful actors with substantial political resources and a capacity to mobilise for policy change. In US biofuels policy, the role of the oil industry was not always prominent in the mid-2000s, but the highly visible campaign led by the API in the last three years suggests a change in strategy (and, increasingly, of policy-makers' responsiveness to that strategy).

Another policy dynamic in conditions of market and technological change, as planned for in US biofuels policy, is that the set of feasible policy alternatives – the set of available policy ideas that could plausibly work – expands. In contrast to much institutional analysis, that suggests options not taken at a point in time tend to recede or diminish with time, policy to support an infant industry, and legislated commitment to rapid expansion in production and consumption of a product premised on technological change, might be a case where policy has the effect of making new policy tools and instruments available over time. In other words, policy-driven market expansion through technological change may also generate new policy instruments. In addition, and separately, key drivers of the development of policy ideas are perceptions of the consequences of existing policy; and the construction of policy problems. If existing policy is perceived to generate adverse outcomes, politicians, bureaucrats and policy experts are more likely to undertake a search for or develop new policy alternatives to address those problems.

This sort of expansion of the policy menu is particularly likely to emerge in political settings in which policy expertise is broadly diffused, like in the US, rather than concentrated in a small, cohesive elite as – arguably – in the EU policy-making system. In the US biofuels case, for instance, the coalition of 'anti' groups against RFS2 has easy access to a diverse system of academic and think-tank experts and to a large set of relatively open institutional venues through which to build coalitions. These include institutions that set a broad societal context (for example, the media – a channel we discuss in more detail in Chapter 9) as well as more detailed policy agendas (for example, legislative committees, government commissions).

In terms of considering the policy menu for US biofuels, it is relatively easy to categorise biofuels policy instruments according to the different stages of the supply chain which they try to affect. Policy instruments are divided into those operating at the input level, the biofuel production or distribution levels, and the consumer level. Production of biomass can be stimulated by subsidies either for the biomass itself or to reduce the costs of production. If an energy feedstock also goes to markets other than fuel (corn for ethanol or for animal feed), then the support can either be limited to the biomass market or be applied to all production.

Production of biofuels may be aided by a range of capital grants, guaranteed loans and tax credits. Biofuel revenues can be enhanced both by subsidies and by protection from foreign competition. There can also be fixed prices for biofuel sales. Subsidies to encourage the distribution of biofuels are also paid in some cases: these can be in the form of tax credits or direct subsidies. They can also entail quantitative obligations to aid biofuel distribution, such as the introduction of pumps that can dispense biofuel. Many biofuel policies rely on supporting the demand for biofuels, through subsidies and tax credits to blenders to supplement the mandates that attempt to control the share of biofuel in blended products.

The US has a raft of farm programmes that support the production of corn and soybeans, with direct payments based on historical entitlements, countercyclical payments that compensate for low prices and loan deficiency payments that give a measure of price support. Subsidised crop insurance is available, along with a several forms of revenue insurance. However, these programmes operate regardless of the destination of the crop – to food, feed or fuel uses.

As previously noted, the US mandate for the use of biofuels is not straightforward in its effect, but as long as the RFS remains in place it will be the crucible in which debates about the future of US biofuels policy are played out. It sets out the superordinate goal towards which policy changes in other instruments are geared. As discussed previously in the chapter, the RFS2 requires that 36 billion gallons of biofuels be used in the US for road transportation, an amount that could account for perhaps one-quarter of all road transport fuel sales by 2022. Ethanol produced from sugarcane is not viewed to be a 'conventional biofuel' under the Act: it can be used to make up the advanced biofuel mandate and (if cost-competitive with other biofuels) the difference between the overall mandate and the advanced mandate. This presents the Brazilian sugarcane ethanol industry as a crucial actor in US biofuels policy.

In addition to support for current production, the US has also initiated programmes for research into biofuels, particularly the development of new forms of biomass that may be cheaper and not compete with the food use of cereals and oilseeds. In the US, public spending on biomass R&D was approximately $800 million over the period 1993–2004, more than eight times the amount spent by the next largest spenders in this area: Japan, the Netherlands and Sweden (Josling *et al.*, 2010).

The operation of the RFS has been subject to voluble arguments both for and against. This provides a rough proxy for the 'winners' and 'losers' from biofuels policy and how this may be shifting over time towards the 2022 targets. Supporters of the RFS claim it serves several important policy goals. Most prominently, the RFS reduces the risk of investing in renewable biofuels by guaranteeing demand for a projected period and helps solves the 'hold up' problem in business–government relations. It enhances US energy security by promoting production of a liquid fuel from renewable, domestic sources. This in turn results in a decreased reliance on imported fossil fuels.

For some, the RFS continues to matter as a boost to the US agricultural sector, by providing an additional source of demand for the US agricultural products, and increases rural incomes and rural employment opportunities. Supporters argue that renewable biofuels go unrecognised for the full extent of their environmental benefits. Unlike gasoline and gasoline additives, biofuels are non-toxic and biodegradable, and emit substantially lower volumes of direct greenhouse gases than fossil fuels when produced, harvested, and processed under the right circumstances.

The RFS has received criticism both from those interests who lost in the original RFS policy design, but also those who perceive their losses to have been greater than expected from the EISA expansion of the original RFS. Opponents argue that, by picking biofuels as a sort of 'winner' alternative fuel, policy-makers may have excluded or slowed down the development of other, potentially preferable, alternative energy sources. This has pushed influential environmental groups more strongly into anti-biofuels political action.

In terms of concrete business interests, some argue that the RFS distorts the market and investment incentives, redirecting venture capital and other investment dollars away from competing alternative energy sources. Instead, these critics have argued for a more 'technology-neutral' policy. This advocacy strategy was employed successfully in 2011 and 2012 against the continuation of large federal support for ethanol production on the grounds that the sector is no longer in its economic infancy and would have been profitable during much of the period 2007 until 2011 without federal subsidies.

For others, RFS2 has had substantial, and in 2007 terms unintended, consequences in other areas of policy importance, including energy/petroleum security, pollutant and greenhouse gas emissions (as seen above), agricultural commodity and food markets, and land-use patterns (as discussed in detail in Chapter 9). Though biofuels are liquid fuels that can be incorporated into the current storage, distribution and pumping infrastructure, the ability to do so has its limits. Biofuels have a number of properties that will make it difficult to fully and effectively incorporate them into the petrol infrastructure and vehicle fleet. Ethanol can differ from petrol in a number of ways, depending on the particular fuel: lower energy content, different physical and combustion characteristics, and corrosive properties. These differences would require modifications to current infrastructure and vehicle design.

The political tensions inherent in the value conflict between rural development, energy security and climate change mitigation have been exacerbated by the longer-term policy effects of RFS2. In the original design of the mandate, there is not a complete set of instruments established to compensate key biofuels interests for the adverse consequences of an uncertain market (given agricultural commodity and oil price volatility) alongside technological uncertainty about the blendwall for first generation ethanol and the feasibility of industrial scale production of advanced biofuels.

It may be self-evident to state, but it is only in an economic environment characterised by high oil prices, technological breakthroughs and a high implicit or actual carbon price would biofuels be cost-competitive with petroleum-based fuels. And in many ways this is the core tension or dilemma with the policy, the element that still functions to undermine confidence in achieving the ambitions of EISA to the full. The paradox is that a shock to energy policy may create, at the outset, a sense of needing to do something amongst both policy-makers and citizens. This leads to a technology-based solution that, on paper, addresses the problem the shock has caused, but in practice is infeasible: 'any rational legislator arguably *should* vote for technological mandates no matter how far-fetched, since they seem to promise the kind of solution voters will view most favourably' (Grossman, 2012: 48, emphasis in original). In the US (and, indeed, in the EU), the promises made about advanced biofuels seem to be falling foul of exactly this internal contradiction.

Postscript: The debate as of early 2014 and the threat to the RFS

As of early 2014, it appears that a number of the threats facing US policy are coming to a head, with the debate encapsulating several of the concerns set out above. Pressure has been building on the RFS in recent years, via the EPAs annual final rule-making on (revised) annual targets for different biofuels within the RFS. Indeed, the target was even set aside in 2012, following a Washington court ruling which agreed with the API, that the EPA mandates were 'unreasonable'. Whilst this has, thus far, had no lasting effect on the RFS and the decisions taken by the EPA, in the latter part of 2013 the EPA changed its framing of the issue, to one based on the blendwall and the ability of the market to accommodate more ethanol.[4] Thus the focus has shifted from the supply of ethanol to the effective demand for ethanol. This provided the backdrop to the EPA's proposed rule for 2014, published in November 2013. This seeks cuts to all but one element of the RFS in 2014 (all figures in billion gallons, taken from an article on the Biofuels Digest website)[5]: cellulosic ethanol (1.75 to 0.017), advanced biofuel (3.75 to 2.2), corn ethanol (14.4 to 13.01), thus renewable fuel (18.15 to 15.21), plus biodiesel (1 to 1.28). Advanced biofuels are thus threatened with a 41.33 per cent cut, corn ethanol with a 9.7 per cent cut.

EPA discourse may have switched to the blendwall, but supporters of biofuels argue that the blendwall was known about in 2007, and that the expectation was that the fuel industry would do something about it: they have not done so and seemingly are now being rewarded for it (in the eyes of those biofuels supporters). Indeed, we were told in one interview back in 2010 that the anti-biofuels interests did not want to address infrastructure issues – especially regarding cellulosics. One thing several interviewees told us was that RFS2 was much more about cellulosics (creating the conditions for their development and thus enabling the crossing of the technology bridge to advanced biofuels) than about corn ethanol which, as it was put in Chapter 1, we can 'do' already without significant policy intervention being required. Brent Erickson has put the issue thus:

> The RFS was intended not to codify that only ten per cent of transportation fuel be biofuel, but instead to actually help break through the blend wall by incentivizing investments in higher blends of ethanol, drop-ins, and biodiesel as well as new advanced biofuels – and it has been working as intended.[6]

It is also suggested (in the 17 November Biofuels Digest article) that, for reasons related to the rising cost of purchasing RINs, the oil industry could get White House support for backtracking on the RFS by pointing to the impact of rising RIN prices on petrol prices for consumers. A separate article on *Biofuels Digest*[7] goes so far as to argue that there is a deliberate policy of over-charging for E85 (comparing ethanol and gasoline price spreads on wholesale and retail markets), reducing demand and thus adding to the blendwall constraint.

Putting this development into the wider context of White House agenda-setting, it is an oddity of US biofuels policy that the big push for biofuels was led by (Republican) President George W. Bush, from an oil state (Texas). It now appears to be under threat from a (Democratic) President Barack Obama, from a corn-belt state (Illinois). However, this apparent paradox is easy to explain when set against the backdrop of the shifting debate over the nature and meaning of energy security in the US. After 2001 President Bush, in successive State of the Union addresses, identified biofuels as a way of helping improve energy (oil) security and reducing oil imports, ramping up the rhetoric until it led to the RFS and RFS2.

What he could not have foreseen was the shale revolution and hydraulic fracturing, as outlined earlier. Moreover, this new source of energy is relatively cheap, and it does not come with the associated bureaucracy of, for example, RINs. Not surprisingly, therefore, the oil and fuel industries have responded completely differently to this than to biofuels and the RFS. In the last three State of the Union addresses, President Obama has been making fewer and fewer references to 'energy'. His references to energy security are now linked to shale deposits, and when discussing renewables he names wind and/or solar, not biofuels.

The pressure on US biofuels policy is manifested through debates over RFS2. The figures shown above indicate a degree of ring-fencing for US corn ethanol, with only biodiesel – which is not subject to a blendwall – likely to be allowed to expand. Given the failure to deliver the technical developments that could have dismantled, or at least moved, the blendwall, the EPA's decision looks increasingly to the pro-biofuels lobby like supporting the position taken by the likes of the oil industry. Perhaps, more seriously, in the longer term, it risks creating significant tensions between corn/first generation ethanol interests and advanced ethanol interests. This could split the pro-biofuels coalition, possibly cementing the corn ethanol element but risking the long-term future of the advanced ethanol industry: it is not only rock that risks being fractured by shale gas exploitation.

5
Comparing Biofuels Policy Drivers – Common Themes, Differences and Issues for Analysis

Introduction

The analysis in this chapter is built around two themes: first, we review the material from Chapters 1–4, to identify common themes, similarities and key differences between the policies of Brazil, the EU and US; second, we introduce briefly the ideas that will form the basis of the more formal analysis contained in Part II of the book.

In the first part of this chapter, we shall see that whilst energy security, rural development and emissions reductions drivers have influenced policies in all three cases, in each case both the meaning and emphasis given to each driver have varied over time in distinct ways. The varying policy histories of our three cases also lead to key policy differences. Notably, given the primary timeframe of our research (the 2000s), we compare directly EU and US policy, with the very different history of Brazilian policy offered as a point of contrast. This is not to deny the EU, and especially the US, policy developments in previous decades, but our particular interest here – and especially in Part II of the book – is the 'biofuels frenzy' of the 21st century. In this, it is important to recognise that, whilst substantive and ongoing ethanol policies have been implemented in Brazil for decades, this more recent period has also seen significant developments, as seen in Chapter 2.[1]

Comparing the basics of our case studies

Perhaps the most obvious first point to make, yet one that is absolutely critical for some of our later analysis, is that in all three cases, there is both ethanol and biodiesel production and consumption – and that, in all three cases, first generation biofuels dominate. There are small-scale localised exceptions to this; for example, domestic UK biofuel production is based almost entirely on waste and recycled inputs – which, as explained in Chapter 1, we are defining here as a second generation biofuel – whilst the US advanced biofuels industry, based on cellulosics, is gradually seeing more commercial plants coming on-line. The aggregate picture, however, is of the dominance of first generation biofuels.[2] The logic behind this was analysed in Chapter 1. First, production of these biofuels utilises existing feedstocks, production technologies and usage technologies (current-technology road vehicle engines). Second, our interviews with senior policy officials in all three cases confirmed the view that first generation biofuels were seen as the basis for establishing biofuel markets, which would then be developed in particular through an expansion of advanced biofuel production and use. The latter require investments and time to develop and to bring to market in sufficient volumes. A market built initially on first generation biofuels could also help determine the potential scope for biofuels. This, supported by appropriate policy frameworks, would encourage and facilitate the investment required to develop and bring to market those advanced biofuels. This is an absolutely fundamental feature of biofuels markets and policies in all three cases: repeatedly, interviewees in all three cases referred to the notion of a 'bridge' – in particular that first generation biofuels represented a 'bridge technology', to a market led by advanced biofuels.

The situation has arisen, however, that the focus on this long-term direction for policy and for biofuel markets has been clouded. A central reason for this is that, whilst everybody involved knew that first generation biofuels were feasible technically and deliverable practically, the full extent of their *potential* downsides was not known initially – especially in the context where production and use were being ramped up to historically unprecedented levels over a very short period of time. This is not to say that future advanced biofuels, of whatever generation, will be without downsides: to recall from Chapter 1 a quote from one interviewee, 'the best biofuel is always the next one, because we do not yet know its downsides', it is just that the full extent of the possible magnitude of the possible downsides of first generation biofuels was not appreciated fully. This has been compounded by the ongoing failure to deliver significant volumes of advanced biofuels to markets: several of our interviewees referred to 'policy getting ahead of the science'. Indeed, one US interviewee in a department of federal government argued that 'from a science perspective we see a lot of inconsistencies'.

Both the EU and US policies clearly did anticipate the existence of potential downsides to first generation biofuels. They incorporate sustainability criteria, thresholds of environmental performance (such as GHG emissions performance relative to fossil fuels), land-use restrictions, incentives to promote the development of advanced biofuels and so on. That said, the efficacy of those regulatory measures has itself also come under scrutiny. With the bifurcation of biofuels policies, between the containment of the potential downsides of first generation biofuels, and the continued promotion of advanced biofuel research and development, this element in the biofuels debate is now questioning the ability of policy both to manage the downsides of first generation biofuels, and to encourage the development and delivery of advanced biofuels to market. A feature of the biofuels debate in recent years has been the emergence and escalation of lobbying against not only 'biofuels done badly' but, increasingly, against the very existence of policies which support first generation biofuels. This pressure is coming, in particular, from development NGOs (for example, ActionAid, Christian Aid and Oxfam). In some cases, the narrative against biofuels either ignores or plays down the differing impacts of different biofuel generations – yet this simply reflects the reality of biofuels production currently.

Thus whilst there is a bifurcation of biofuels policies, between first generation and advanced biofuels, one particular challenge for policymakers and for the biofuels industry is that this debate appears to be taking place in a political environment where the discourse of the anti-biofuels interests fails to make a clear distinction between first and advanced generations, and their potentially very different impacts. That said, as another of our US interviewees said, 'advanced biofuels are five years away ... and have been five years away for thirty years'. With advanced biofuels, when will the hypothetical finally become real, on a scale sufficient to enable them to take over and drive biofuels markets? And, from the point of view of the anti-biofuels lobby, how much damage will have been done in the meantime?

As of early 2014, the evidence outlined above seems to suggest that, in the US, the RFS is coming under rapidly increasing pressure. In the EU, the reform proposal seeks to take policy a step away from the support of first generation biofuels. That said, in April 2014 the Commission published guidelines for how member states should provide financial support for renewable technologies, to comply with EU state aid rules. These permit the continued support ('operating aid') for facilities already producing first generation biofuels, up to 2020. They do not, however, permit support for new facilities producing first generation biofuels ('investment aid').[3]

In contrast to the EU and US, in Brazil the reaction to pressures around the production and use of first generation biofuels has been not just sanguine, but positive, in particular with reference to their dominant biofuel – sugarcane ethanol. Interviews with several of the most senior officials in Brasilia who work on biofuels issues revealed a common position: this scrutiny has been nothing but a good thing for Brazil. Specifically, as concerns have been raised over first generation biofuels and their emissions performance, land-use change effects and so on, officials have gone back, re-analysed their data and found repeatedly that Brazil's ethanol production stands up extremely well to even the closest scrutiny. Indeed, several interviewees in Brasilia referred to Brazil's sugarcane ethanol as a first generation biofuel with advanced biofuel performance.

It is worth reiterating at this point that there is nothing intrinsic about first generation biofuels that guarantees their having significant negative impacts – to return again to that Brazilian 'mantra', there is no such thing as good biofuels and bad biofuels, just biofuels done well and done badly. Each biofuel needs to be judged on its own merits, in terms of the feedstock used, the production method used for the feedstock, how it is converted into biofuel, the technology pathway used for this, the domestic agricultural context of the feedstock production, competing products and markets of those feedstocks, the scale of any production of feedstock and biofuel and so on. Against this complex backdrop, Brazilian sugarcane ethanol delivers everything expected *and hoped for* from a biofuel. Even so, where ambiguities remain in terms of side effects, there remains debate – as we discuss further in Chapter 9.

The challenge, explored below and through the rest of the book, is getting policy-makers elsewhere to be willing to accept that not all biofuels can deliver such performance and that, as a result, policies need to ensure that only 'biofuels done well' are promoted. The difficulty is that, unlike Brazil, other countries' biofuels markets are still emerging (even those of the EU and US). As a result, a conflict can arise between policy-makers' desire to keep promoting first generation biofuels in order to establish the markets, and those lobbying against biofuels (especially first generation), who are concerned not only by the downsides of first generation biofuels, but the lack of progress on commercialising advanced biofuels, a failing which is helping perpetuate the market dominance of those first generation biofuels.

Comparing the use of the three policy drivers as arguments for biofuels policy

The preceding chapters have demonstrated clearly that all three drivers have featured in the determination of our three focus biofuels policies. Rather than repeat what has already been analysed previously, our purpose here is to consider further how those concepts have been utilised. In so doing, we illustrate how the concepts are sufficiently broad and flexible to accommodate the different specific uses to which they have been put by policy-makers, in different countries, in different contexts, at different times.

As recorded in Chapter 3, one of our interviewees in Brussels, a senior European Commission official in a key Directorate-General for biofuels policy, but one who was critical of the direction EU biofuels was taking, referred to the three policy drivers as legs of a tripod, but one where the policy did not rely on all three to support it. Thus, biofuels policy could be defended against criticism of any one of those legs. For example, if a biofuel was questioned in terms of its capacity to deliver reduced GHG emissions, it could be defended if it substituted for fossil fuel, and/or if it helped create or sustain rural employment. We can now take this idea further, and argue that the very nature of each 'leg' is not fixed, that there exist – potentially – multiple definitions of 'energy security', 'emissions reductions' and 'rural development', utilised as required by policy-makers to support a given biofuel and a given policy action.

Energy security

The first oil crisis, of 1973–1974, had a huge economic impact globally. Chapter 2 outlined the consequences of this for Brazil as a country that, at the time, imported more or less all of its oil requirements. As a result, a major driver behind Proálcool was the desire to be less dependent on energy – especially oil – imports. Proálcool was, moreover, part of a multifaceted drive for renewable energies, which also promoted huge investments in hydro-electricity. In the intervening years, however, Brazil has discovered and begun to exploit, in particular, significant pre-salt oil deposits offshore (the Campos Basin was discovered in 1974, as Proálcool was being prepared and which, in one go, doubled Brazil's known oil reserves). Even so, nearly one-half of all Brazil's energy needs is supplied from renewable sources. Because of the earlier investment in hydro-electricity, about 80 per cent of electricity is generated from renewable sources, about 85 per cent of which is from hydro (currently, about 5 per cent is generated from biomass, including sugarcane bagasse, as discussed in Chapter 2).

An interview with key officials in Brasilia revealed that Brazil's government, in the 1970s, decided to continue with Proálcool, but also to provide Petrobras with the incentives to continue seeking new oil reserves. As those interviewees put it, '30 years later we are harvesting the benefits of that decision, with two simultaneous energy policies which benefit the country a lot'. The growth in total energy demand, driven by sustained economic growth, has been well-served by the relatively balanced growth in the supply of domestic renewable and non-renewable energy sources.

As for whether the exploitation of the oil reserves was seen as undermining Brazil's green energy credentials, the officials were quite clear that this was not the case. Under existing energy policy law, there is a commitment to increase the share of renewable energy, whilst recognising that the world will continue using oil for decades to come. Thus, in the context of global oil demand and a growing domestic economy, Brazil's energy mix is not operating as a zero-sum game. As a result, the country is able to exploit its fossil fuel reserves, whilst continuing to promote green energy.

Brazil has, since 2004, been more or less self-sufficient in oil – but the internal mix varies between petrol, of which Brazil is an exporter, and diesel, which Brazil continues to import. Interestingly, one interview also revealed that, just as Prodiesel was designed, in part, to promote wider regional development efforts, so too has been the exploitation of these oil reserves. Thus whilst refineries in the South East of the country supply oil and oil derivatives for the domestic economy, those in the North East are the base from which those products can be exported. Overall, therefore, since the 1970s energy security has been central to Brazil's renewable energy policy developments, including those for biofuels. Moreover, even as oil reserves and domestic production of oil derivatives has grown, helping to promote domestic energy security in the sense of self-sufficiency, successive governments have taken a wider, more strategic, view of what energy security can mean. Energy sources have diversified, whilst the exploitation of those energy sources is being put to wider domestic economic and social purposes.

The US case, however, has seen energy security defined primarily in terms of reducing the country's dependence on imported (essentially the Middle Eastern) oil. In the 1970s, the first oil crisis caused particular problems for the US economy as it had become extremely oil-dependent, based on decades of ready access to cheap oil from domestic reserves and, increasingly, from the Middle East, where five of the so-called Seven Sisters were US-owned.[4] On occasions, US policy has even sought to go beyond energy security per se, instead aiming for energy *independence*. This latter term, which has reappeared in parts of the US policy discourse in recent years, was first coined by President Richard Nixon in the immediate aftermath of the first oil price hike in 1973: 'What I have called Project Independence 1980 is a series of plans and goals set to insure that by the end of this decade, Americans will not have to rely on any source of energy beyond our own.'[5]

This proved an impossible target, but one consequence of the oil crisis and the massive rise in world oil prices was that it enabled more oil reserves, in more parts of the world, to be exploited. This was simply because reserves that were harder to access than those in the Middle East, and thus much more expensive to access, now became economically viable (Chalabi, 2010). This resulted in a falling share of world oil markets for the countries of OPEC over time. Indeed, Chalabi argues convincingly that higher oil prices were actively supported by the Seven Sisters, who would benefit from the higher oil prices and from new non-OPEC investment opportunities. He also argues that higher prices were supported by the US, as a critical means by which not only could new (more expensive) non-OPEC (and non-Soviet Union) oil reserves be exploited; they would also help make nuclear, gas and renewable energy investments viable.[6]

By the time we visited Washington in the Spring of 2010 to conduct our first set of interviews, advertisements on buses and bus stops showed barrels of oil, piles of dollars and the image of Iran's President Ahmadi Nejad, accompanied by the slogan 'Iran makes a KILLING every day we wait: It's time for action on clean energy + climate'.[7]

In the EU, meanwhile, energy security has taken a different meaning again, in the context of its biofuels policy. As shown in Chapter 3, reference is made to 'traditional' notions of energy security throughout the policy documentation published since the turn of the century. That said, if EU member states are to increase biofuels use to any great degree at all – let alone by enough to get even close to a 10 per cent blend – either the biofuels or the feedstocks with which to make biofuels will have to be imported. Thus, as interviews with senior officials in the European Commission confirmed, in the EU the notion of energy security involves diversifying fuel types and source countries. Moreover, whilst EU policy does not distinguish explicitly between fuel types, it is significant that domestic production of biodiesel is greater than that of ethanol. In terms of energy security, this is important, given the EU's net imports of diesel, but net exports of petrol. It should be added that one of our interviewees indicated that Brazil's notion of energy security also embraces the idea of diversifying sources – a position entirely in keeping with their goal of seeing a global market develop for biofuels.

Rural development

Rural development is a concept prevalent in many countries' agricultural support policies. It is a concept that is, also, defined in different ways in different contexts and in different countries. Often, the definition extends beyond agriculture, to include the wider rural economy. This broad framing of the term is reflected in its use as a biofuel policy driver. In the EU, rural development policy is a multidimensional policy, added to the Common Agricultural Policy (CAP) as Pillar II in 1999 (see Ackrill, 2000). It operates on a seven-year programming cycle, with the member states agreeing a ceiling on available spending from the EU budget, then choosing from a long menu of varied policy options (the menu altering in each cycle). In the European Commission's Directorate-General for Agriculture and Rural Development (DG-AGRI), biofuels policy is located under Rural Development.

One interviewee, a senior Commission official who played a leading role in developing key aspects of the RED, argued that biofuels policy should be promoted primarily for what he saw as its intrinsic energy-related benefits – that it was capable of standing on its own merit in this regard. He recognised that it was very helpful, secondarily, to point to the policy providing economic opportunities in rural areas, but that this was not of primary design concern. Rather, it was a result of having a policy designed for other reasons but which, given the industry's largely rural location, would generate these wider benefits.

One issue addressed during our interviews in Brussels was whether there was a causal link between CAP reforms and the development of biofuels policy: that as the CAP was reformed and liberalised, was biofuels policy a back-door way of providing indirect support to farmers? More specifically, was there a link between sugar policy reform and biofuels policy development? Our sceptical interviewee within the Commission had no doubt. In the Summer of 2005, following the political agreement on EU sugar policy reform (November 2004) and the publication of the final (and slightly altered) legal texts in February 2005, discussions between the Agriculture Commissioner and Commission President led to a Cabinet level working group on biofuels, to facilitate the sugar reforms by showing that biofuels could provide a significant market outlet. This interviewee argued that a key part of the discussion was then to ensure biofuels were differentiated by the GHG emissions performance. Another flashpoint was to minimise the unintended environmental consequences of biofuels production. Both were finally incorporated in the resulting text, to some extent.

Several other interviewees, however, flatly denied any link between CAP reform and the development of EU biofuels policy. What surprised us as long-standing researchers of the CAP was the extent to which the interview with members of commodity teams within DG-AGRI revealed the total embedding of the principle of policy de-coupling, the process by which, as far as possible, the link should be broken between policy support levels and farmers' production. Indeed, this was one reason why the Energy Premium, introduced in the 2003 CAP Reform to promote the production of inputs for renewable energy, was abolished in the 2009 'Health Check' reform – as it was incompatible with de-coupling (although one interviewee also admitted that it was found to be rather an ineffective instrument). We are also unable to triangulate any links between CAP reform and biofuels policy with our previous analysis of the 2005 EU sugar reform (Ackrill and Kay, 2009; 2011a). In no interview for that project did anybody make a connection between sugar and biofuels policies. That said, our sceptical (biofuels project) interviewee's statements cannot easily be dismissed, given his insider-status to key discussions. Currently, therefore, the link between CAP reform (especially sugar policy reform) and EU biofuel policy remains to be confirmed or refuted categorically.

In the US, 'alternative energy' is one of Rural Development's 'Seven Strategies for Economic Development'.[8] That said, multiple interviewees in Washington, with officials in government departments, Senate and House staffers, and industry bodies, were all clear that US ethanol policy is about 'Big Ag'...not 'rural development' in the broader sense. Chapter 4 has shown just how close these links between ethanol and agricultural (corn) support policies actually are. Indeed, these links date back to the 1970s, well before the MTBE/oxygenate debate.

One important element of the US debate is the extent to which biofuels, and ethanol in particular, generate rural jobs. As interviews with government department staff and staffers to House and Senate Committees made clear, ethanol tends to be a bipartisan issue, the split being determined more by whether you come from a corn state or not – especially if the latter happens to be an oil state. As with sugar policy, it is not just the agricultural commodity itself that creates such a powerful lobby, but the combined forces of the wider related industrial interests located along the supply chain. It is, however, worth reflecting on the fact that it was President George W. Bush, from Texas, who drove the push for renewable energy after 2001. Thus, against such a dramatic backdrop as 9/11, energy security was a sufficiently powerful factor that a president from an oil state pushed so hard and so effectively for renewables.

Brazil, once again, is a case apart. Since the economic liberalisation of the 1990s, most of its markets have essentially been free. Agriculture and related economic activities are undertaken primarily on the basis of market considerations. This is helped by the scale and efficiency of Brazilian agriculture. It is no coincidence then that the poorer parts of the country, such as the North East, are also places where agriculture is less efficient. Here, as explained in Chapter 2, over many decades there has been considerable (over)-representation in Congress, helping to ensure that these parts of the country receive assistance in various forms. As explained in Chapter 2, and above in this chapter, the ethanol, biodiesel and even oil industries have been utilised in successive governments' efforts to help promote economic development in these regions. Perhaps, in a large, emerging economy like Brazil, the notion of rural development is trickier to nail down than the likes of the EU and US, as rich, urbanised, (post)-industrial economies. Instead, a primary concern for politicians is trying to ensure less unbalanced regional growth and economic development. Rural development is promoted not so much by a distinct set of rural development policy instruments (as the EU has), but by the government utilising various policy levers, by which it can use all sectors and industries in the promotion of regional economic development more broadly.

Climate change mitigation

This is, arguably, the most controversial aspect of biofuels and biofuels policies. To the extent that biofuels are promoted as one part of an energy matrix that is being rebalanced, away from fossil fuels, they must deliver reduced GHG emissions compared with those fossil fuels. Their ability to do this, however, is contested, for a variety of reasons. In this chapter, we outline the key policy differences that have arisen in each of our three cases.

Both the EU and US have sustainability criteria, which seek to ensure that the biofuels entering the transport fuel mix are delivering GHG emissions benefits, by excluding 'biofuels done badly'. The extent to which each country puts an emphasis on this policy driver, however, does vary considerably. As Chapters 3 and 4 showed, environmental and climate change considerations have played a much greater role in the development of EU biofuels policy than the US. Indeed, we argued in Chapter 3 that EU biofuels policy emerged from a decade of environmental and climate change policy developments.

Abstracting from the debate over the efficacy of sustainability criteria, we can see differences between the EU and US expressed in the different targets and thresholds for biofuels deemed sustainable, under the RED and FQD, and under the RFS. In the US, the RFS to 2015 is dominated by conventional ethanol which, in practical terms, means corn-starch ethanol. The threshold for this is a 20 per cent GHG emissions reduction compared with fossil fuels. Higher emissions savings are demanded of biodiesel and advanced ethanol: 60 per cent reductions (cellulosic ethanol), or 50 per cent reductions (all other advanced biofuels). Moreover, the latter can include the likes of Brazilian sugarcane ethanol.

In the EU, there is no distinction between different types of biofuel, insofar as GHG emissions performance is concerned. The initial minimum emissions reduction compared with fossil fuels must be at least 35 per cent. This is higher than the 20 per cent minimum in the US for the (still) dominant biofuel, corn ethanol. That said, in the US other forms of ethanol, as well as biodiesel, must from the outset deliver a higher emissions reduction – either 50 per cent or 60 per cent. In the EU, under the original policy, the threshold only rises to these levels in 2017 and 2018. Since these figures will apply to all biofuels in the EU, by this point EU policy will have 'overtaken' US policy, because the latter includes the grandfathering of older ethanol production facilities, plants that are not required to improve the GHG emissions performance of their ethanol over time.

The actual consequences of this are extremely hard to determine. The US mandate, of 36 billion gallons of biofuels by 2022, is higher than (perhaps double) the EU's ceiling of 10 per cent of transport fuel – not all of which even has to come from biofuels. That said, the current expectation is that neither target will be met. In the EU, the total biofuel figure for 2020 is unknown, other than that it is expected to constitute the dominant part of the 10 per cent renewable energy figure. Moreover, raising the minimum GHG emissions threshold above 35 per cent, whether it happens as originally scheduled or as set out in the reform proposals, will have a major impact, because it will rule out potentially significant volumes of biofuels, especially biodiesel, acceptable currently under the 35 per cent emissions savings target.

In the US, meanwhile, regardless of the difficulties being faced currently, longer-term challenges remain. The scope for producing and incorporating the mandated volumes of advanced biofuels, between 2015 and 2022, into the transport fuel mix as laid down in the RFS appears currently to remain a distant dream. Thus, with conventional ethanol production grandfathered, unless there is a significant substitution of advanced for conventional ethanol within a more limited RFS, the scope for biofuels delivering *significant* future emissions reductions in the US, beyond the 20 per cent (minimum) delivered by corn ethanol, looks modest.

In Brazil, the performance of sugarcane ethanol, as discussed above, is such that this country has never developed sustainability criteria in the style of the EU or the US. That said, it has recognised potential land-use issues, by introducing an extensive and detailed agro-ecological zoning scheme for new sugarcane production, and a similar but less rigorous scheme for oil palm. With the exception of this last development, biodiesel production also has no particular environmental constraints imposed on it. These were felt unnecessary, given that 'the general environmental laws of Brazil apply and that biodiesel does not need to comply with any additional or specific requirements'. (Stattman *et al.*, 2013: 28; Garcez and Vianna, 2009 offer a critical review of Prodiesel's environmental policy omissions.)

Recent developments in ethanol trade, however, raise questions about the overall emissions performance of the biofuels used in Brazil. For several years, Brazil was able to produce enough ethanol to maintain the 25 per cent blend in petrol, as well as maintain adequate supplies of E100, to give flex-fuel vehicle drivers complete choice over their fuel mix. In recent years, however, there have been poor cane harvests, high sugar prices and volatile oil prices – as a result of which Brazil's ethanol production has fluctuated. The upshot of this has been that, at times, there has not been sufficient ethanol to supply the domestic market. The Brazilian government has, on occasions, reduced the mandated ethanol blend in petrol, to 20 per cent. Meanwhile, reduced supplies of E100 have driven up prices, thus reducing demand. These events have combined to generate an odd pattern of ethanol trade.

In the US, large areas of policy-making are devolved to the individual states. The California Air Resources Board (CARB) has set a performance target for biofuels based on carbon savings that US-produced corn ethanol typically does not deliver, but which Brazilian sugarcane ethanol does. Imports of Brazilian sugarcane ethanol into California have, however, on occasions left the Brazilian market short of ethanol to blend into petrol – as a result of which, in recent years Brazil has imported varying volumes of US corn-ethanol. This circular trade, at first glance, looks silly, however impeccable the (policy) explanation behind it. But not only does the transportation of this ethanol result in additional emissions, the de facto replacement of domestic ethanol with US corn-ethanol in the Brazilian transport fuel mix worsens its per unit volume emissions performance. Another factor making Brazil's transport emissions picture worse still is that whilst more or less all Brazil-produced cars have flex-fuel engines, economic growth has seen a large rise in the number of vehicles being imported. They are incapable of being run on E100 or a petrol-E100 mix, but only the petrol blended to 20 per cent or 25 per cent ethanol. Thus, other things being equal, this further reduces the per-vehicle use of ethanol relative to petrol in Brazil, resulting in a commensurately worse emissions performance.

The three drivers – Mutually reinforcing or mutually contradictory?

Given the foregoing analysis, the answer to this question is, inevitably, 'it depends'. We summarise these links in Figure 5.1. One striking feature of this is that the only unambiguously positive relationship is from rural development to energy security (that is, promoting rural development via biofuels will have a positive impact on energy security). The reverse relationship is either positive if the feedstock is produced domestically; zero otherwise (this latter case could represent a boost to rural development, narrowly defined, in other countries, but we recognise this is a highly complex issue, especially insofar as it involves developing countries).

The relationship from energy security or rural development to climate change mitigation depends on a number of factors, for example technology pathways for conversion; but in Figure 5.1 we identify what we consider to be the main determining factor – the feedstock used. In short, biofuels done well will see climate change mitigation enhanced, but biofuels done badly will see climate change mitigation undermined. The reverse relationships, from climate change to energy security or rural development, are, in this simplified analysis, assumed to depend on policy responses to the initial relationships. That is, if energy security or rural development concerns lead to biofuels produced from feedstocks which undermine climate change goals, what then happens?

If policy-makers continue with those particular biofuels, the relationship between the variables is negative. The continued production of

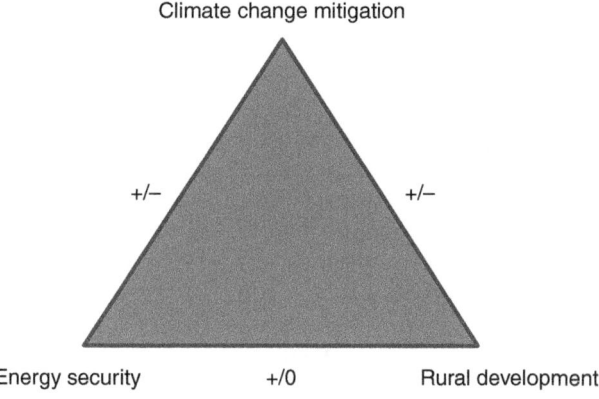

	To:	Climate change	Energy security	Rural development
From:				
Climate change		- - - - - - - - - -	+/−: Does use of the'wrong' feedstock result in a reduction in biofuels production or not?	+/−: Does use of the 'wrong' feedstock result in a reduction in biofuels production or not?
Energy security		+/−: Depends on nature of the feedstock	- - - - - - - - - -	+/0: Depends on whether we refer to domestic rural development
Rural development		+/−: Depends on nature of the feedstock	+:Domestic production will boost energy security	- - - - - - - - - -

Figure 5.1   Links between the three biofuels policy drivers

those biofuels will continue to enhance energy security or rural development. If, however, policy-makers respond by reducing or eliminating production of those particular biofuels, the relationship will be negative, in that energy security and rural development would be undermined, in order to promote climate change mitigation. This negative relationship can, however, be avoided by replacing those biofuels which undermine climate change efforts with biofuels which enhance it: promoting either or both of rural development would, under this scenario, also promote climate change mitigation.

Conclusions

In this chapter, we have explored how the three key drivers of biofuels policy have been utilised in our three cases. Notably, this has included a consideration of how policy-makers have exploited the flexibility of interpretation available to them regarding what each of the drivers actually represents. A simple reading of this allows us to draw the following conclusions. First, to the extent that biofuels substitute for fossil fuels, and given that there is not congruence between source countries of fossil fuels and biofuels, energy security is enhanced. This applies in all three policy cases, in the context of the differing definitions of energy security adopted in each. Second, given the varying definitions of rural development present in our three cases, biofuels help promote that policy goal in all three cases, in a variety of ways.

A cynic might stop us at this point and argue that we are simply defining energy security and rural development in whatever way necessary to ensure that biofuels are shown to benefit those policy goals. To this we would reply that it is not us, but the policy-makers, who are doing that. It is also at this point, however, that this easy assessment breaks down. The third driver of biofuels policy, climate change mitigation, cannot be so easily dealt with. Quite the opposite – it is this dimension of biofuels policy that gives rise to more or less all of the most divisive and contested aspects of biofuels. Some of these arise directly from an analysis of the climate mitigation goal, for example, whether or not the life-cycle emissions of a given biofuel are even below those of the fossil fuel it is intended to replace, let alone by how much. Other concerns, for example, the impact of biofuels on food prices, arise from the interconnectedness of biofuel, agricultural commodity, food product and oil markets. The sustainability criteria used to control the downsides of first generation biofuels – or, more accurately, the production of the feedstocks for first generation biofuels – themselves represent the source of another policy concern – the extent to which biofuels policies are, or are not, WTO-compliant.

It is thus with these concerns in mind that Part II of the book focuses its detailed analysis of biofuels policy around related key themes: of policy design challenges; policy implementation challenges; specific challenges presented by the international dimensions of policy; and a range of specific challenges identified in Part I: ILUC, food v. fuel and consumer/media framing of biofuels debates.

Part II

6
The Challenge of Policy Capacity in Biofuels Policy Design[1]

Introduction

A feature of biofuels policies that we have been reminded of in writing this book is the sheer ambition of policy-makers in seeking to establish and drive the rapid and considerable expansion of biofuels production and use. Yet this ambition has another dimension to it: are policy-makers in the 21st century capable of such direction of markets? The scale of policy-makers' ambition runs counter to several decades of debates and literatures on government overload, the shift from government to governance, the hollowed out state and government failure. These concepts reflect changing economic and political relationships, not only domestically but also internationally, as trade barriers come down and economic interdependencies – via increasingly global supply chains – expand. Against this theoretical and empirical background, the capacity of policy-makers to deliver on their promises over biofuels cannot be taken for granted.

The concept of policy capacity is variously defined across several fields and enjoys currency because it encompasses two linked governance concerns: (i) the recognition of increasingly complex interdependence in the international economy imposing limits on conventional command-and-control governing strategies and (ii) a prominent discourse about the development of steering capacity as an essential feature of effective management of the increased governance complexity that inevitably results from the changes outlined above. A critical dimension of this challenge, as the rise to prominence of the concept of governance over the last two decades attests, is that governments are being required increasingly to engage with non-governmental and international organisations in making and implementing policies in new ways. This requires the state to adopt a new role – that of partner in a network of actors, rather than as leader in a hierarchy.

The biofuels policy processes from Part I are powerful examples of the policy capacity challenge. The EU, US and Brazil all highlight – in different ways – the policy-maker's dilemma of maintaining longer-term policy goals as credible commitments, even though considerable flexibility and adaptability in policy-making is required to reach those far horizon goals, in conditions of high technological and market uncertainty. In such terms, biofuels policy is always an intertemporal choice which tests the capacity to account for the future benefits of a low carbon future in current policy processes, because if the pathway to their achievement is uncertain and politically contested in the implementation phase (and Part I shows they certainly are), then those future benefits from biofuels policy may be heavily discounted, shortening policy-maker horizons and rendering the overall transition process politically vulnerable. Brazil offers an important contrast: whilst EU and US policies seem confounded by this dilemma Brazil, in an earlier period, had developed a complex mix of policy instruments to manage it.

Policy capacity is thus not exclusively a matter of analytical capabilities; governments are called upon both to make the right policy choices in design but also to implement them effectively. Design and implementation have clear linkages, and in this chapter we focus primarily on challenges in policy design, but also introduce implementation. We then look at domestic implementation in the face of specific external pressures in Chapter 7. The initial part of this chapter sets out a basic framework for the analysis of policy capacity using five core dimensions. This framework is used in the main body of the chapter to analyse the policy roads to biofuels in the EU, US and Brazil, drawing on the content of Chapters 1 to 5. The challenge presented is manifest: the costs and benefits of biofuels policy-making are both spread out over extended periods of time, making the politics of coalition-building amongst diverse groups complex and prone to disruptions.

Policy capacity for the transition to a low carbon economy

The salience of, and challenge to, the ambition to build policy capacity is related to the oft-invoked view that governments 'steer and do not row' in the contemporary international economy. Parsons (2004: 44) develops a nautical version of the steering metaphor, arguing that map-making and navigating are core elements of the capacity to govern, 'the ability to chart the voyage, plot coordinates, set direction and take a long term view – in short, to navigate'. The academic literature on policy capacity offers subtle definitional variety. An indicative selection includes the ability of governments to make intelligent choices (for example, Painter and Pierre, 2005); the ability to scan the horizon and set strategic directions (Howlett and Lindquist, 2004); the faculty to weigh and assess the implications of policy alternatives (Bakvis, 2000), as well as aptitude in making the effective use of appropriate knowledge in policy-making (Parsons, 2004; Bakvis and Aucoin, 2005).

For others, it is the ability of governments to decide upon *and* implement preferred choices of action that characterises policy capacity. Parsons (2004) describes this as the 'weaving' function of modern governments. Given the range of non-government actors involved in the policy process, governments must demonstrate the ability to weave together the multiplicity of these organisations and interests to form a coherent policy fabric, which is robust enough to survive pressures on the politics of both policy design and implementation, over extended periods of time.

To ease exposition, we have avoided a systematic literature review on policy capacity in this chapter and instead extract a framework with five specific dimensions from the literature. Although far from exhausting all the possible dimensions of policy capacity, the framework is employed to dissect analytically and specify policy capacity in our three cases, to understand how they seek to effect a transition to a biofuels economy. As we shall see below, policy capacity to effect this transition is challenged not only through changes to governments' ability to govern, as described by assorted metaphors above, but also as a result of a feature of biofuels described in Part I of the book – that biofuels policy is complicated by the market for biofuels overlapping multiple other markets, including those for individual feedstocks, for food, for oil and so on.

Value agreement capacity

This dimension refers to the role of governments in facilitating agreement about the values that motivate policy. This dimension contributes to policy capacity by enabling governments to introduce significant policy changes without risking societal backlash and electoral defeat. This feature of policy capacity is related to what is sometimes called integrative capacity (Howlett and Rayner, 2006b) as well as to the ambitions for policy integration prominent in the fields of environmental studies. However, as contributions in Adger and Jordan (2009) suggest, this may be an unrealistic standard in many areas of the governance of sustainability, which are marked by conflicting values and sharply divided public attitudes. Here the capacity of policy-makers is better understood in terms of managing the conflicts inevitably arising from the incommensurable values involved in the transition to a low carbon transport fuel mix.

Selection capacity

Following the literature, this dimension is the varying ability of governments to forge authoritative choices which commit relevant governmental and social actors, notably private enterprises in the case of biofuels, to implementing policy alternatives. As the environmental policy literature suggests, the goal of sustainability imposes a set of requirements across a wide set of cognate policy areas. This leaves governments with the challenge of coherent, joined-up policy-making in a context where power may be diffuse, political consensus difficult to achieve and implementation requires strong coordination between multiple policy and market activities.

Operational capacity

A key proposition emerging from the network governance literature is that open and inclusive policy networks, although fluid in membership and often difficult to institutionalise, actually contribute to effective policy-making because they are able to absorb complexity and can be more resourceful and resilient in delivering outcomes than closed and exclusive networks (see, for example, Bovens *et al.*, 2001; Koppenjan and Klijn, 2004). Policy-makers trying to effect a transition to a low carbon economy must deal with political, technological and market uncertainties that require they act within existing sector-specific networks, as well as across different networks that are being linked by novel biofuel policy agendas.

Foresight capacity

The ability of governments to look forward and anticipate is often doubted, and the attribution of myopia brought about by electoral timetables or media cycles is common. However, even amidst pervasive uncertainty, effective policy-making requires a capacity to identify 'known unknowns' in order to map a future path for policy. Foresight is a distinct dimension of policy capacity; for analytical purposes, there is a relevant difference between unconscious and conscious incompetence amongst governments. Understanding the capacity to anticipate potential political as well as private enterprise 'hold up' problems is important, independent of evidence of the ability to resolve them.

Reflection capacity

Even if the long-term policy goal (defined as part of a government's Foresight Capacity) is fixed, Reflection – as a map-reading exercise – requires governments to monitor on an ongoing basis where they have come from, where they are going and how they might get there. This dimension encompasses the faculty for detecting problems with current policy settings, as well as the ability to learn both cross-nationally and longitudinally, in order to respond to evidence of failings, or deviations from the long-term path. The capacity of policy-makers to avoid mere 'muddling through' behaviour when faced with existing policies identified as ineffective and/or unpopular is an important element of the steering mechanism available to governments. The extent to which the policy-making system as a whole has access to, and can utilise, institutions that allow current policies to be critically examined, to look back to recover historical lessons and scan the international horizon for alternative policy change options is constitutive of policy capacity.

Policy strategies governing the transition to a biofuels economy

As Part I of the book has outlined, in both the EU and US over the last decade, biofuels policy has been transformed from a relatively minor, niche aspect of energy policy into a central, economy-wide plank of future energy strategy in the transport sector. Despite their dissimilarities as political systems and the dissimilar constellation of political interests in sustainability policy, both jurisdictions arrived at a shared objective, at roughly the same time, of promoting a dramatic increase in the production and consumption of biofuels in the liquid transport fuel market. In contrast, in response to volatility in oil and sugar prices in the 1970s Brazil has displayed different dimensions of policy capacity in developing a distinctive and sophisticated mix of policy instruments in the biofuels domain over a 20-year period through to the mid-1990s. This policy capacity has facilitated the liberalisation reforms described in Chapter 2 and increased significantly domestic production and consumption. Brazil has realised the objectives of increasing energy security and promoting rural development, via a dominant biofuel delivering significant GHG emissions reductions relative to fossil fuel, in a more advanced and coherent manner than the EU and US.

The rest of this chapter employs our framework of policy capacity to understand the processes of policy design in the EU, US and Brazil that lead, at different times, to this common and long-term objective, as well as their experiences in policy implementation. This includes the processes of error adaptation, learning, new political alliance formations, as well as shifts in the broader policy environment which have all contributed to test the policy capacity of governments to design and implement biofuels policy.

Policy design

As Chapters 3 and 4 showed, biofuels policy initiatives in the EU and US from the 1970s remained largely a niche concern but they did lead, eventually, in the EU to the 2003 Biofuels Directive, and in the US to the Energy Policy Act of 2005. These, in turn, laid the groundwork for enactment of the key salient policy episodes, the policies in situ throughout the period of research for and writing of this book, the US 2007 Energy Independence and Security Act (EISA) and the EU's 2009 Renewable Energy Directive (RED). In both cases, the US and EU introduced two different types of policy to satisfy the incommensurable values in the policy process. Along with market, political and technological uncertainties, different values ranging from energy security to sustainability have been accommodated by adopting a policy design in which there was a demarcation between first generation and advanced biofuels. These differences reflect the different dimensions of first generation and advanced biofuels as laid out in Chapter 1.

Over a much longer period, Brazil has had two broad policy phases. The first, under the Proálcool policy, used a mix of policy instruments to provide the incentives for fuel retailers, consumers and car-makers to demand and supply ethanol in the volumes required to establish a mass market. This phase was sufficiently successful, in ethanol but not yet biodiesel, to act as a bridge for a second phase of policy reform: the set of liberalisation measures enacted that has allowed Brazil to establish almost a free market in ethanol. As previously described, biodiesel policy was designed much later in the mid-2000s but seems to promise the potential for a similar two-phase reform trajectory.

Ethanol policy, it should not be forgotten, took nearly 30 years to go from the establishment of Proálcool to a free market, with FFVs giving drivers freedom of fuel choice at the pump. We analysed in Chapter 2 how the Brazilian government is taking a slow and steady approach to Prodiesel, recognising that production capacity needs to be developed, but also that constraints remain currently to the full integration of small-scale farmers into the policy through the Social Fuel Certificate. Thus effective policy design relies on recognising the long-term nature of key aspects of policy capacity, as elaborated on below.

Policy implementation

Although designed to promote domestic policy objectives, the biofuels policy environment has attracted the attention of non-domestic political actors in the implementation phase: international institutions like the OECD, international research institutes such as IFPRI with a focus on trade and development and advocacy groups of non-state actors organised cross-nationally. This was manifest in the politics around the world food price spike in 2008 – and the resulting international opposition to biofuels – which came as biofuels production was accelerating – as shown in Figures 1.1 and 1.2.

There were also specific market factors for expanding ethanol demand in the US, related to Methyl Tertiary Butyl Ether, as analysed in Chapter 4. It is also worth noting that this take-off followed a period of oil price rises: from the start of 2003 to the middle of 2006, crude oil prices rose by 2.5 times, falling back to (just) double January 2003 levels at the start of 2007, rising to 4.5 times January 2003 levels by the middle of 2008. The higher the price of oil, the more likely biofuels are to be cost competitive, even without subsidies, and the higher the financial burden of oil imports. This economic dimension of energy security added to the range of arguments set out in Part I.

The international scrutiny of biofuels policies is related to their entanglement with many different societal values (security, environmental protection, economic development and rural communities). The unprecedented media scrutiny over the first seven months of 2008 to the possible role of biofuels in the global South's food security crisis brought international attention to biofuels policy in all three cases, and shaped the implementation politics by directing attention to the impact of biofuel production on food security and the heightened incentives to convert lands in the developing south to biofuels production. This sits alongside environmental sustainability, with various attempts (from both inside and outside of government) to model the net energy benefit of biofuels over their life cycle versus conventional fossil fuels, including the direct and indirect land-use impacts of biofuels on GHG emissions (see Chapter 7 for how EU policy, in particular, has resisted these pressures, and Chapter 9 for further, detailed analysis).

Although these models are subject to controversy, their impact is to suggest to policy-makers that greater environmental gains will come from advanced biofuels, whose GHG savings are larger, and which avoid the food-fuel trade-off (although the possible land-use change effects of producing some biomass sources remain contested – as discussed further in Chapter 9). This reinforces the panaceaic properties assigned to advanced biofuels; they perform the function of separating incommensurable values in the policy process, assuaging opposition to first generation biofuels by promising that any harmful effects will be dissipated by technological change in the advanced biofuels policy. This is an important element in the deliberate tolerance of ambiguity of policy goals, in combination with the conscious design of market and technological uncertainty into the policy process.

Assessment of policy capacity in the transition to a biofuels economy

Value agreement capacity

The development of biofuels policies in the EU and US has been underpinned by reference to common factors, albeit with varying emphasis reflecting different domestic political constituencies, as analysed in detail in Chapter 5 (see also Gamborg *et al.*, 2014). Of the three dominant policy drivers – energy security, climate change mitigation and rural development – in the US the greater emphasis initially was given to energy security, but in both the details of RFS2 and in the politics of the ILUC debate, environmental concerns are manifest. A policy to reduce imports of Middle Eastern oil has had particular resonance in the context of 9/11 and the subsequent military engagements in Afghanistan and Iraq, as explored in detail in Chapter 4. EU policy has long-referenced energy security but, with limited domestic production capacity relative to longer-term mandates, the emphasis has been more on diversifying energy imports, in terms of both fuel types and source countries.

In Brazil, for both ethanol and biodiesel, the stress in the original policy design was on energy security and rural development whilst, over time, it has also adapted, in particular, its policy rhetoric to the international agenda of climate change mitigation. This, in the language of historical institutionalism, is evidence of policy conversion: policy institutions remain the same but are reinterpreted, re-enacted, so as to *'redirect to new goals, functions or purposes'* (Streeck and Thelen, 2005: 28, emphasis in original). Thus, as international attention on climate change grew, Brazil was able to bring its existing ethanol policy into its wider set of policy responses to emissions and climate change concerns.[2]

Whilst US policy referenced climate change mitigation in the period up until 2010, the global economic crisis and the shifting federal budget situation have seen this value decline substantially in salience in the reforms of 2012 described in Chapter 4. In the EU, however, this has always had greater emphasis, especially given the role the EU has given itself as a global leader in climate change politics and policies, as analysed in Chapter 3. Rural development has also played a notable role in the EU and US although, as we saw in Chapter 5, in the US biofuels policy has been seen more overtly as an aspect of 'traditional' agricultural policy, rather than rural development per se.[3]

The political tensions inherent in the value conflict between rural development, energy security and climate change mitigation have been exacerbated by a commercial environment characterised by market uncertainty (of the price of oil, commodity prices, and so on.) alongside extreme technological uncertainty about the blendwall for first generation ethanol in the US, and the feasibility of industrial scale production of advanced biofuels, in the US and EU – but not in Brazil where, as one of our interviewees put it, Brazil's sugarcane ethanol is a first generation biofuel with advanced biofuel performance.

Thacher and Rein (2004) provide a typology of practical solutions to managing value conflicts in policy-making (see also Stewart, 2009). The US and EU cases reveal a novel variant of the structural separation or firewalling strategy, where value conflict is mitigated through organisational design: different public organisations involved in a policy are guardians of different values. In the biofuel cases, the novelty exists in policy designed to separate conflicting values intertemporally. In both the US and EU, there are short- to medium-term goals of increasing the production of first generation biofuels, alongside longer-term goals for advanced biofuels production. In Brazil, in contrast, value conflicts were not evident in the design and implementation of Proálcool. Instead, the values of energy security and rural development interacted symbiotically as positive drivers of policy reform trajectory of the 1980s and 90s, with this symbiosis maintained as ethanol was brought into the climate change mitigation agenda.

Politically, the advanced biofuels policy element in the US and EU functions as a superordinate policy goal to describe the overall direction of travel (see also Foresight Capacity). This helps to persuade those concerned with values of food security and ecological sustainability to tolerate large increases in first generation biofuel production – which use agricultural feedstocks, with associated land-use changes and food production effects – as a necessary bridging step to an advanced biofuels economy. Reflecting this intertemporal aspect of policy, mandates increase certainty for those investing in advanced biofuels technologies, but also seek to commit future policy-makers to the values (and policies) determined in the present.

The two-part policy design enacted in both the EU and US has had to evolve in implementation as a result of endogenous tensions associated with technological and market uncertainties as well as, since 2008, significant shifts in the broader policy environment. Neither the EU nor the US have solved the problem of reconciling the ambiguity of policy values by the first generation/advanced biofuels design separation. The novel requirements of environmental sustainability are firmly part of the first generation policy debate, rather than being confined to the advanced biofuels element. This is seen in both the EISA and RED, which set out how to calculate the net energy benefits of biofuels on a life-cycle basis.

The main uncertainties remaining in the first generation part of biofuels policy involve the complexities of understanding and modelling land-use change effects, especially ILUC, and the possible impact of biofuels production on food prices. These, analysed in detail in Chapter 9, remain intensely debated aspects of biofuels production and policy. Perhaps an even greater challenge to biofuels policies overall is the principal uncertainty in the advanced biofuels element of biofuels policy – when will those advanced biofuels, with their panacean properties, finally come to market? As well as representing an ongoing policy failure in its own right, delays in the development of commercial advanced biofuels at scale only reinforce the belief that we shall be stuck with first generation biofuels for some time to come. This, in turn, brings even more attention to the issues of ILUC and food v. fuel surrounding first generation biofuels (see also Söderberg and Eckerberg, 2013).

The key design feature of policy in both the EU and US in accommodating ambiguity in policy goals, through a sequence of temporal targets to guide the transition to mass biofuels consumption, alongside profound market and technological uncertainties about the feasibility of the targets, has thus set the conditions for implementation politics. A central feature of policy in these jurisdictions has thus been that the demand side policy was enacted *without* coordination with supply availability and import security concerns. Sustainability and land-use change concerns were incorporated into the EISA, but in the EU whilst the RED outlined sustainability, the details were agreed and published over a year later, and, as seen in Chapter 3, as of early 2014 the ILUC debate is ongoing in terms of how it might be reflected in policy. In Brazil, meanwhile, the shift to a free market achieved with the introduction of FFVs largely obviates the need for policy-makers to coordinate the demand and supply sides of the market. We say 'largely', because over the period since 2009, shortfalls in anhydrous ethanol production have led not only to (market-driven) imports, notably from the US, but also the policy decision to reduce temporarily the mandatory ethanol blend in petrol, from 25 to 20 per cent.

Selection capacity

Recent work by Patashnik (2008) on the political dynamics of major reforms offers insights into the factors which explain why some reforms stick and others are unpicked. In particular, reforms that endure through the implementation process typically destroy an existing policy subsystem and reconfigure the political dynamics. This insight is highly relevant to the ability to commit societal actors to a policy direction, or selection capacity in our terms. At least in terms of its political salience, biofuels is a new policy domain in the EU and US, and overlaps with several existing policy domains with their own particular networks and legacies. The policy capacity challenge is to be able to reorganise these in the direction of achieving long-term mandates for biofuels use.

President Bush used 9/11 as the background to his assertion that the US needed to reduce its dependence on imported oil. The 2005 Energy Policy Act, partly in response to changes in the policy on oxygenates in petrol, introduced the RFS to create a usage target for ethanol. This was followed, two years later, by a greatly increased usage mandate in the EISA, introduced via RFS2. In the EU, 2001 saw the publication of proposals for what would become the 2003 Biofuels Directive. This had complex roots in ongoing policy discourses on sustainable development, the Lisbon Strategy and the embedding of environmental considerations into a range of common EU policies. Initial voluntary biofuels usage targets were replaced in 2009 with mandated, and much greater, usage targets via the RED. In both jurisdictions, however, expanded mandates were to be delivered over a longer timeframe. Mandates, as already noted, sought to increase investor certainty and seek to bind future policy-makers. This would be consistent with Parsons' notion of 'weaving', discussed above.

The conflicts between interests/actors in the biofuels processes under scrutiny were not resolved in a single policy act, but rather have continued beyond adoption or enactment. The two-part biofuel policy design (first generation/advanced) enacted in both the US and EU was a strategic move to commit to – or 'select' – a path in a dynamic policy system rather than a one-off choice in some classic textbook way. The promise or foreshadowing of a radically changed policy environment, notably a mass market and international industry in biofuels, provided policy logic in the system, favouring some interests or values over others. Nevertheless, this selection has not determined uniquely the subsequent policy path, which instead remains open and contingent to shifts in policy settings which arise from political competition and technological change. In Brazil, although there is a clear continuity in ethanol policy across three decades, the liberalisation reforms are related to changing policy preferences about economy-wide economic policy, energy policy as well as the widespread introduction of FFVs after 2003. Moreover, lessons from ethanol were utilised in the design of core features of biodiesel policy – in particular the wish to embed small family farmers at the heart of the policy, as a counterpoint to the dominance of large-scale agri-business in the sugarcane-ethanol industry.

There are important intertemporal dimensions in the EU and US in terms of selection capacity. The strategy has been to agree policy objectives in the short term, and accommodate different and conflicting values involved in biofuels expansion by legislating guides to future policy change in the advanced biofuels sector. This adoption of two separate types of biofuels policy has precipitated a distinctive implementation situation. There is a superordinate goal of a substantial biofuel component in the liquid transport fuel market but there is only limited explicit policy guidance over the sources of future advanced biofuels. There also exist complementary superordinate constraints such as sustainability criteria which, as noted, are perceived to constrain first generation biofuels more than advanced.

In the EU, additional 'credit' against renewables targets is available for advanced biofuels. In the US, the RFS includes separate targets for different types of advanced biofuel, but this is accompanied by powers granted to the EPA to amend those targets year on year, in line with expected production. Uncertainty over what industry actors are willing and able to deliver raises the question of whether, and how, this acts as a constraint on the evolution of policy, in which actors are committed to a certain policy direction, a concern reinforced, in the US case, by the inclusion in the policy implementation process of this reverse gear.

Our interview data shows that reformers' attention has waned in the US and EU, that the initial spirit that held the biofuels policy coalition together has been contested by novel, unanticipated concerns in the implementation process. Even in the case of the mature, well-established, Brazilian ethanol policy, adjustment has been necessary in response to the dilemma of reconciling ambitions for national development involving the efficient use of all of Brazil's energy resources, a free-market in ethanol and an international policy environment in which reductions in GHG emissions is stressed prominently.

Both the US and EU cases reveal that the discourse of sustainability, far from helping to integrate various policy objectives in the implementation phase in a common concern, has instead generated significant problems. It can add complexity to initial policy design; increase the number of objectives, interests and players in policy implementation; and can be used as a key resource by institutionally embedded actors in existing policy legacies, to protect their autonomy against the ambitions for a coordinated implementation of the initial biofuels policy design. This is complicated further by the continued absence of significant commercial-scale production of advanced biofuels from the market.

Of course policy is always made under conditions of uncertainty, but the degree of market uncertainty and technological uncertainty that attends biofuels policy is central to selection capacity. For example, ethanol in the US is a valid outlet for its large-scale corn production, whilst ethanol and biodiesel are valid outlets for EU agricultural production, in the context of decoupled support under the reformed Common Agricultural Policy. These are 'market' reasons for this use of US/EU feedstocks that exist independently of 'policy' reasons. Biofuels policy implementation faces particular challenges with uncertainty. It must contend with volatility in both agricultural commodity and oil markets, competition domestically and internationally that is variable, all in the context of a credit crunch for capital. The upshot is that, in trying to create appropriate incentives for a new biofuels industry, biofuels policy settings are not the only relevant factor.

Operational capacity

The design of biofuels policy in all three jurisdictions requires flexibility and adaptation in operation, to ensure success. In the US, first generation ethanol is restricted by the blendwall. Advanced biofuels are limited by the lack of domestic commercial production – although that target could be met through imports of, for example, Brazilian sugarcane-based ethanol. The policy response to the blendwall has been to seek agreement on raising the ethanol blend percentage in petrol, whilst, for advanced (cellulosic) biofuels, the current response is for the EPA to reduce the annual targets in RFS2, in line with estimated production each year. Meanwhile the Department of Energy (DoE) is offering increasing amounts of money to support research on different advanced biofuels, including the establishment of demonstration-scale plants. There are already several million FFVs on the road in the US but, as detailed in Chapter 4, this is more to do with tax breaks and thus lower purchase prices, whilst the availability of E85 remains limited and somewhat localised.

The EU mandate target of 10 per cent of transport fuel from renewables in 2020 is somewhat more modest than the US target, perhaps half of the US figure, based on current projections for US petrol demand for 2022 (although the environmental impact differential may well be less than this, given the greater GHG emissions reductions demanded in EU policy). Also, the mandate does not distinguish between ethanol and biodiesel, nor between first generation and advanced biofuels (offering both flexibility in delivery and reducing potential ethanol blendwall concerns), and – crucially – imports are expected. The EU target, as a result appears, in theory, to be more achievable. Countries' progress towards their individual targets, however, remains very mixed, as Figure 3.1 showed.

Given the increasingly binding nature of the blendwall in the US, a significant exportable surplus of ethanol is emerging from what was intended primarily as a domestic policy. Furthermore Brazil has, on occasions recently, imported significant volumes of ethanol, given poor recent cane harvests and price developments on the sugar market. These imports have therefore come, in part, from the US (see also Chapter 9). It is also the case that, in the US, ethanol exports do not count towards the RFS2 mandate. At the time EU policy was being developed, the lack of policy coordination was such that the understanding that imports would be needed was not matched by knowledge of where those imports were to come from.

Beyond trade concerns, our data on the scope of the networks in which the EU and US are actively involved casts doubt on the extent to which domestic biofuels policies are subject to the influence of transnational politics. The principal intergovernmental biofuel body is the Global Bioenergy Partnership (GBEP). Interview data reveal very little transnational political impact on domestic policies in this forum. Instead, for example, it provides a forum for countries to 'learn about' each other's jurisdictions, policies, the repertoires of instruments and methods, rather than representing an expert community that crosses international borders. The national politics are so competitive it seems that this 'soft' supranational power is attenuated and does not redistribute existing power in the policy system.

A notable difference between the US and EU experience does emerge, however, as a result of the different trade needs arising out of their domestic biofuel usage mandates. Our interview data from policymakers in the US reveals very little evidence of a direct transmission belt from the ideas, strategies and activities of transnational political actors to domestic politics. In contrast EU policy, with its need for imports, has had to face a series of significant transnational issues. Given that GHG emissions reductions is a central part of EU policy, there is a need to ensure the production of all biofuels feedstocks contributing to biofuels consumed in the EU avoids conflict with this goal. As a result, a system of policy measures is being developed regarding the production conditions of feedstocks, whilst remaining consistent with WTO rules on trade barriers.[4] There is strong evidence that EU sustainability criteria have been developed with WTO rules expressly in mind (see Box 3.1). Moreover, transnational political engagement is expressly part of the guidance offered to countries in the WTO Technical Barriers to Trade Agreement (TBTA) when developing such criteria (these issues are discussed in detail in Chapter 8).

There are also various national and international Roundtables, in which a variety of economic and civil society actors work together in defining what sustainable biofuels are or should be, what could/should private actors do to move in that direction and how this can be certified and verified. Direct engagement between governments and Roundtables is key where the practical requirement of aligning certification schemes of the latter with importer government standards sets the biofuels trade policy agenda. As explored further in Chapter 8, the EU, in many cases, utilises existing networks of certification bodies. Their work with their existing standards is helping promote globally the sustainable production of a range of products, independent of biofuels concerns.

This allows biofuels to be embedded into an established notion of 'sustainable production' – albeit via a certification system that is not without attendant challenges, as analysed in Chapter 8.

The differences between EU and US engagement with such transnational actors reflect the different trade implications of biofuels mandates. In Brazil, meanwhile, the government's position has been to engage with the EU during the process establishing legislation on sustainability criteria. Brazil has avowedly not, however, supported or promoted domestic producers' involvement with certification schemes. It has remained neutral, a position from which (interview data reveal) it can best monitor policy implementation to ensure WTO compliance in the policy-enactment phase. If it judges it necessary, it can then bring an action in the WTO without risking the credibility of the case: supporting Brazilian producers' involvement in a scheme could give the impression it was giving official approval to that scheme.

Overall, therefore, we have found evidence of transnational political networking activity by policy-makers, but in political terms the influence of this work is attenuated, certainly relative to domestic networks and their impact on domestic policies. Identifying a lack of operating capacity at the transnational scale, particularly in political terms in the form of intergovernmental influence, is an important finding: biofuels are being traded internationally but, other than with the EU sustainability criteria, there is little evidence of coordinated policy engagement at the international level. One interpretation is that at the present stage of the development of biofuels policy, the construction of domestic operating capacity remains of paramount importance. Moreover we may assume that, a priori, this is less challenging politically than constructing operational capacity for policy on a transnational scale. As noted earlier, Value Agreement was undertaken by means of emphasising aspects of biofuels policy to which domestic constituencies would be most receptive.

In considering the coordinating mechanisms to implement the policy in each jurisdiction and how this has left both policy areas susceptible to fragmentation through sustainability concerns (which, in turn, have different meanings and scope in the EU and US), the initial policy design is brought into analytical focus. This included various policy objectives – implicitly and explicitly – such as rural development, energy security and GHG emissions reduction, which were managed intertemporally. This strategy required the bridge between the two stages, of first generation and advanced biofuels, to be credible for coordinated implementation. Without this, spillovers between the different temporal dimensions of the policy may arise, and values conflict as their incidence is distributed amongst different policy legacies that might have an interest in biofuels policy.

It may be self-evident to assert that the lack of coordination in policy design is positively related to lack of coordination in policy implementation. Yet this truism sits at the heart of the implementation dilemma for both the US and EU: the need to remain credible about long-term goals as established in the policy design, alongside ambiguity about the reasons for those goals and necessary uncertainty about the steps to their achievement.

One policy issue that has been used by opponents of biofuels to challenge the policy goals, that raises questions about the internal consistency of domestic biofuels policies, and has exposed significant differences of opinion in different countries, is ILUC. Introduced in Chapter 1, we analyse ILUC in detail in Chapter 9. Given the causal complexities intrinsic to ILUC, and its unobservability, it can only be estimated – and any model and its estimates are limited by our understanding of this complex concept – understanding which is evolving continuously. This illustrates how policy-making within the discourse of sustainability is never a settled process because of its expansive quality; different and novel dimensions constantly emerge, compared to something more limited (but nonetheless grand) like GHG reduction, whilst understanding and modelling capacity of these aspects continues to develop.

Foresight capacity

To adapt a distinction in March (1978), policy-making always takes place in an environment of uncertainty about the consequences of action, but importantly also in an environment in which there is uncertainty about future preferences. The study of policy processes, over many decades, has served to query, empirically, the use of the assumption that policy preferences are stable over time, consistent and exogenous to policy actions and their effects in modelling the policy process. Instead preferences are often vague, unstable and inconsistent.

With both the EISA and RED, there are explicit biofuels mandates (plus other targets, such as the US Corporate Average Fuel Economy, or CAFÉ,[5] target for increasing average car fleet fuel efficiency). The EISA also contains explicit targets for the improved energy efficiency of federal car fleets, as well as support for research into new technologies such as electric cars and advanced biofuels. In the EU, such measures are not contained in the RED, but in a range of thematic research pro-grammes, as well as other components of the 2009 Climate and Energy Package of legislation, outlined in Chapter 3. In addition, there are pol-icy measures that are less explicit, yet continue to move energy usage and climate-change mitigation measures in a particular direction. This latter feature is not only a direct legislative effort, but also represents a contribution to a wider political discourse on energy and climate issues that continues to be contested.

As the EU Climate and Energy Package shows, we should not for-get that whilst we are focused on transport biofuels, this discussion is located within a wider energy and climate context. In the EU, there are the 20–20–20 targets, discussed in Chapter 3. That said, whilst the EU package includes mandates for domestic energy consumption coming from renewable sources, and improvements in energy efficiency, in the US the Senate removed a House plan to require that, by 2020, electric-ity utilities delivered a minimum 15 per cent of electricity derived from renewable sources, with up to 4 percentage points able to come from efficiency savings (Sissine, 2007: 2).

In Brazil, as analysed in detail in Chapter 2, the institutional struc-ture of policy management has changed to reflect the broader policy context beyond biofuels. In addition to CIMA, which deals specifically with ethanol, the CNPE was established to take a broader perspective on the energy market. The lead department is the Ministry of Mines and Energy, but it includes several ministries, NGOs and academics, and it reports to the president in its role as responsible for proposing energy policies. MAPA joined only in 2008, given its role with biofuels. As elaborated in Chapter 2, interviews by the authors with several senior government officials in Brasilia indicated this confirmed ethanol as an energy policy, a shift from its earlier status as an agricultural policy. This reflects an ambition to bring greater coherence to the future policy pro-cess: there was not one group in agriculture trying to increase their share and income, but rather they were operating in the broader context of energy.

In the EU, the institutional structure allows the European Commis-sion, not subject to electoral cycles, to engage in more challenging 'blue-sky' policy thinking. As early as 2011, it began discussions about possible goals for energy policy by 2050.[6] However, even if, in policy terms, the EU is working towards more challenging targets, and even if the discussions of policy in 2050 are not fully delivered, in both juris-dictions the legislation and the surrounding discussion are seeking to take policy in a certain direction. Even if it is two steps forwards, one step back, as the House/Senate disagreement over renewable electricity illustrated, that is still a net step forwards in a particular desired direction.

Reflection capacity

Given the problems facing policy-makers, the question arises of the ability of policy-makers to reflect on policies as implemented. Can inconsistencies between policy-as-designed and policy-as-implemented be identified and corrected? Can lessons be learned from past experiences and/or from experiences in other countries? In terms of the biofuels cases, it is the credibility of the government's commitment to longer-term targets for biofuels use which lies at the heart of implementation politics; changes may be needed in order to maintain forward movement in the general direction of broad goals, even if they are implicit and unquantified. This requires a capacity for forward-looking Reflection.

Our interview data reveal that the feedback processes from policy design to subsequent policy changes are not straightforward; they are an uncertain and volatile mixture of positive and negative. There is evidence of policy layering (Streeck and Thelen, 2005; Mahoney and Thelen, 2010) in design; for example, biofuels policy clearly overlaps with, though it is much broader in scope than, the agricultural policy domain. It increases demand for certain commodities as feedstocks, with biofuels production/refining generally taking place also in rural areas (biofuel feedstocks are typically bulk-losing commodities).

But policy layering in design needs to be complemented by analysis of its consequences in implementation. In particular, the interaction between different legacies and new interests can be observed in the lack of institutionalisation and coordination between the two parts of the design in the implementation phase. The biofuels cases are suggestive of the particular difficulties of joining up government in the new governance of sustainability. The two part policy designs that were enacted in the US and EU immediately introduced a number of bureaucratic players in implementation. The Brazilian case over a longer period reflects the struggle of managing institutionally entrenched agricultural interests within the broader context of energy and resources policy. In the EU several Commission Directorates-General were involved: Energy (DG-ENER), Mobility & Transport, Environment, Climate Action and Agriculture; in the US there was DoE, Department of Agriculture, EPA and State – let alone the other stakeholders and interested non-state actors involved in implementing policy. Furthermore, as noted, the emergence of new international organisations, such as GBEP, provides a forum for discussion between the major players in a more constructive and progressive manner than the traditional international intergovernmental organisation. Internally, however, the biofuels cases have revealed old interests have still been influential. With the RED, DG-ENER held the pen, whilst in the US, RFS2 was driven through by the Senate.

In both the EU and US, despite the use of policy mandates, we observe a biofuel policy bifurcation between an increasingly path-dependent first generation component and a still incoherent, uncertain and contested advanced component. In terms of technological uncertainty and policy implementation, the biofuels policy case is affected by uncertainty about parameters within existing technologies (for example, blendwalls and viability of existing liquid transportation fuel infrastructure) as well as uncertainty about technological paradigm shifts to advanced biofuels. The potential consequence of this latter shift – something anticipated in existing policy design – for biofuels policy-making is so profound as to make its prospect something that adversely affects the institutionalisation of advanced biofuels policy.

The US corn-ethanol production target for 2015 is likely to be met, even if the blendwall prevents all 15 billion gallons from being incorporated into transport fuel, without significant disruption to existing fuel distribution infrastructure (although, as discussed in Chapter 9, the impact of this on food prices is disputed). The growth in the biofuels market up to 2022, however, is designed to be in advanced, especially cellulosic, biofuels. In the US, the different mandates in a sense represent different biofuels policies. So, with the technological change hypothesis, with first generation biofuels policy there is less technological uncertainty and a more coordinated policy. Yet even here, technological uncertainties remain. The blendwall debate has led to a policy for E15 which, without indemnity insurance for petrol suppliers, has yet to deliver much E15 to the market.

Furthermore, technological uncertainties also play a significant role on the demand side. As detailed elsewhere, higher fuel prices have reduced vehicle miles and thus fuel demand, as have CAFÉ regulations. These developments make it harder to absorb even the targeted 15 billion gallons of conventional ethanol by 2015 whilst, as interviews revealed, different options being discussed in response – so-called mid-level blends higher than E15, FFVs and E85 – all have technological implications, including infrastructure and engine-design limitations.

To this can be added the growing unpopularity of biofuels, analysed in Chapter 9.

If the authorities are this uncertain about existing technologies when they make policy related to first generation biofuels, what does that say about policies related to advanced biofuels? It is in the space for these and the huge mandates for a domestic industry not yet working (because technology is uncertain) that the coordination problems are yet greater. Of course, the coordination problems may well be fundamentally technical in nature rather than policy based/political, but the concern of this chapter is how technological uncertainty is related to policy coordination in implementation.

In the EU, the whole context of domestic production of biofuels means imports are more inevitable than in the US, whilst locating biofuels firmly in the climate change agenda has decoupled biofuels policy from agricultural policy. This meant there was less policy layering and the problems which that might pose for coordination in implementation, as the tense relationship between different layers prevents the emergence or design of institutions.

Conclusions

In this chapter, we have developed a basic framework within which policy capacity can be analysed. We have identified five distinct elements within the process of policy-making, its design, reform, implementation and evaluation. Whilst policy-makers in the EU and US wrestle with the dilemma of implementing a policy with far horizon policy goals but which is politically contested in the short- to medium term, so Brazil over three decades provides a contrast case of how biofuels policy capacity can be developed.

Biofuels policies in the EU and US have developed significantly, at about the same time during the last decade. They responded to similar concerns – although across and within jurisdictions, the emphasis given to each individual factor has varied over time. As we have traced their development, significant differences and notable similarities have been observed. In the implementation phase, land-use change, sustainability food security concerns (exogenous to policy designs in both jurisdictions) have pushed both US and EU policy in a convergent direction, with each other but also with Brazilian policy narratives, if not yet policy designs.

We are cautious of such clear-cut analysis, as the chapter simultaneously argues that both jurisdictions are wrestling with the difficulty of coordination and struggling for clear direction in implementation, thus making the detection of any policy trajectory hazardous. Nevertheless, the intriguing point remains that the EU debates (in contrast to the US) have always been located in a wider climate change strategy, right back to the policy design phase. This is an important difference between these two cases. Indeed, we can go further and see separate policy frames at the design phase in the US and EU in terms of deeper cultural values: EU desire to be a 'green' diplomatic leader in the world during the time of the Bush administration and the assertion of deep and distinctive 'European' values, as does the energy security concern in the US tapping into older notions of isolationism, self-reliance and, in particular, energy independence. Yet, the continued contestation of biofuel policy in the implementation phase in both the US and EU especially during 2008 can be observed in increasingly common problems of policy coordination, particularly in managing the 'bridge' between the two types of biofuel policy, with attendant consequences for the politics of implementation. Brazil, meanwhile, is essentially free of the challenges pertaining to the first generation/advanced biofuel technology bridge that casts such a shadow across US and EU policies. Instead, it has seen a policy that, in the 1970s, had more similarities with US concerns over energy security, evolve into one that, with an emphasis on environmental and climate change performance, aligns much more closely with EU policy.

That said, the policy agendas in implementation look more similar in 2014 than they did in the policy design phase, 2006–2007. For example, the tripod of climate change mitigation, energy security and rural development exists in policy narratives in both the US and EU. There is convergence and coherence around those elements as regards first generation biofuel policy. Moreover, as our interview data from Brussels reveal – if someone challenges biofuels with reference to one leg of the tripod – the defence can be made that biofuels deliver one or both of the other two. The continued feasibility of such value-cycling will be tested in both jurisdictions in the near to medium future. As revealed through our interviews Brazilian policy-makers, in contrast, are pretty confident that their biofuels policies stand firmly on three legs, all of which can withstand close scrutiny.

In the EU, the phased introduction of policy has seen the introduction of the RED in 2009, sustainability criteria in 2010, and discussion over land-use change from 2011. With the publication in 2012 of reform proposals, to include an ILUC factor, the Reflection Capacity is being tested to ensure the member states, individually and collectively, reach the stated goals in 2020 – even if, between now and then, further policy reform is needed. Whilst EU policy is sufficiently flexible that its biofuels target can be met, albeit mainly with first generation biofuels, the collapse of reform negotiations in late 2013 casts doubt on how close to the 2020 goals policy can attain. That said, with discussions about policy through to 2030 and even 2050, failure to deliver on the 2020 targets will not, of itself, demonstrate policy failure as such – but rather, the missing of a staging post on an even longer policy trajectory: 2020 can be seen merely as a medium-term intermediate target. In the US, the failure to develop significant production of advanced biofuels, especially cellulosic, represents a serious threat to the ability of the US industry to produce the desired volumes of biofuel types by 2022, whilst the blendwall and the limited availability of higher ethanol-blend petrol and cars to use it represents a serious limit to market demand. Here, Reflection Capacity will be tested fundamentally – and may require a response which questions the entire biofuels policy capacity of the US, in a way that is not the case, yet, with the EU.

7
Biofuels Policy Design and External Implementation Challenges[1]

Introduction

The era of globalisation, as elucidated in Chapter 6, is characterised by widespread claims of a spatial transformation of governance from national to international and transnational scales. This chapter examines biofuels policy-making in these multi-scalar terms, examining the intersection of domestic policy processes and the international arena in the implementation of biofuels policy. We argue that whilst none of the three cases of biofuels policy processes in this book can be presented as operating exclusively within their respective national boundaries, biofuels policy in the EU, US and Brazil is an important example of the constraining power of domestic policy designs manifest even across multi-scalar policy-making structures.

The chapter first introduces the idea of policy layering in domestic design as a means of insulating policy implementation against unexpected effects from the international policy-making arena. Next we draw on the evidence presented earlier about the implementation of the EU, US and Brazilian biofuels policies, to explore this argument. The main domestic policy drivers in all three cases behind the major biofuels policy reforms – energy security, rural development and climate change mitigation – are discussed in terms of policy layers. These are cases of major energy policy reform in an open economy sector that stand analytically for policy design against 'outside-in' linkages. In essence, domestic policy designs use dynamics of policy layering as a form of insulation against disruption in implementation by international level factors and developments. A final section explores how and why layering in biofuels policy design has contributed to keeping it intact, despite manifold pressures in implementation. In this chapter, we identify the principal pressures as concern over GHG emissions (especially those trig-gered by indirect land-use change), and biofuels' possible impacts on food prices and food security. In this chapter, we analyse these as sources of disruption to domestic policy. In Chapter 9, we analyse ILUC and food price issues in much more detail.

We start from the position that the domestic level of policy-making remains preeminent in cases of major energy policy reform. Although we find variation in the effects of the international arena of policy-making across the three cases, this is related to the nature of the domestic policy design. In particular, layering in domestic policy design that nests within a wider policy paradigm, defined in terms of the three drivers of biofuels policy introduced above, may act as insulation against international level disruption mechanisms to preserve the status quo ante: disruption of biofuels policy may require disruption of these broader goals of energy, environmental and rural/regional policies.

Biofuels as a distinctive policy design: Using layering as insulation

Coherence is highly prized in the study of public policy design. The absence of a 'policy glue' (May *et al.*, 2005) holding together goals and objectives is judged evidence of a fragmented policy process and regarded as a failure of rationality in the policy design, leading in turn to implementation problems and diminished policy effectiveness. In the work of Schneider and Ingram (1997) on the symbolic elements of policy design, designs can serve to 'signal' the value of goals and objec-tives, as well as assumptions about target populations, human behaviour and the nature of the problem. From this perspective, a lack of pol-icy coherence for a given policy area may introduce noise into the signals of the design, in particular, signals to different groups in the pol-icy process of the importance placed by policy-makers on their causes and values.

In judging the desirability of coherence in policy design, it is useful to consider the relationship between coherence and the stability of any policy design. Coherence is an equilibrium concept, and the equilibrium in any policy design might be stable or unstable. There is no neces-sary relationship between coherence and the stability of a policy design; this section develops the proposition that in some contexts and under certain conditions, incoherence in policy design does not undermine policy stability. Policy layering is a mechanism which provides protec-tive insulation for the policy design in its implementation. The domestic policy is given stability as it is insulated from potentially disruptive external influences from the international arena of policy-making.

In terms of the relevance and applicability of these ideas to biofuels policies, we have already seen that biofuels policies are responding to multiple drivers, whilst seeking to accommodate the incommensurable values of different interest groups. Despite this, thus far the policies in question have proven remarkably resilient to the inevitable resulting tensions. That said, a clear message from Chapters 3 and 4 is that, in both the EU and US, the pressures building on both policies in 2014 represent an unprecedented level of threat to this stability: a critical juncture, in the language of the academic literature on historical institutionalism.[2]

In broad brush terms, the policy design literature stresses that complexity of policy problems begets complexity in policy design. In other words, there is a strong positive relationship between increasing complexity of policy problems and the required sophistication in the mixes of policy instruments employed to address them (Howlett and Rayner, 2006a; 2007; Howlett, 2009). In contrast, in the literatures on historical institutionalism the role of design is moot; for some, complex policy structures emerge as a result of historical contingencies which are then self-reinforcing, whereas for others complex design is possible as a 'one-off' at a certain point in time that locks-in options and trajectories for policy. The historical institutionalism literatures, however, stress endogenous change and evolving and coevolving institutions, via mechanisms of conversion, adaptation, drift and, the particular focus here, layering (Streeck and Thelen, 2005).

Nevertheless, a role for intelligent policy design is not precluded in historical institutionalism. Equally, the literatures on policy design do not make or require the assumption that policy is made tabula rasa; in Richard Rose's memorable phrase, policy-makers are inheritors more than they are choosers. The aphorism of policy designers may well be: 'I wouldn't start from here'; but policy is still able to be designed using combinations of legacies. In practice, policy design is always constrained, about making choices in context (May, 1992). For historical institutionalism scholars, context includes, importantly, historical legacies of previous policy decisions. The second generation of historical institutional literature (Mahoney and Thelen, 2010) stresses the claim that policy-makers are not institutional dopes, that there is a role for creative agency and that endogenous mechanisms of institutional change can be directed by policy-makers for particular purposes. Indeed, the recently minted concepts of conversion and adaptation in this literature are clearly premised on this possibility, but policy layering can also be seen in such terms.

Policy layering can be an intentional choice, and an element in the repertoire of policy design. Policy layers are intentionally added, as well as extant policy layers intentionally not deleted. Layering may seem antithetical to coherence in policy design because it potentially introduces contradictions between the different layers and sets off subsequent tensions that may require more layers. However, layering may also, simultaneously, contribute to design stability and the extent to which implementation problems emerging from the international arena can be managed. In this case, policy layering is a form of design with the conscious creation of insulation against international level effects on implementation, using layers of policy. Layering in design may, even at the risk of incoherence, provide insulation and resilience against external disruption by anchoring policy in a wider set of domestic policy frames.

In order to elaborate the claim of policy layering in domestic design acting as insulation that constrains the influence of international policy processes, the substantial two level games literature that has emerged from the original work of Putnam (1988) is a useful starting point. Here the general presumption is that international policy processes reallocate political resources in favour of domestic governments, by changing the context in which domestic policy is made. Executives participate directly and play at the two levels, as a result, they are simultaneously able to use tightened domestic constraints to increase their international bargaining power, and use international negotiations to create transnational 'executive cartels' in order to boost domestic control of policy agenda.

On this view of the international-domestic policy intersection, domestic governments are able to exert a degree of control to ameliorate the disruptive potential of information and ideas at the international level of policy-making. In the biofuels policy processes set out below, an established domestic policy design has been confronted by information and ideas mechanisms from the international level. The case studies reveal clear mechanisms that might have upset the policy design, but both EU and US policy has proved stable in implementation so far, in terms of the information about GHG emissions and ideas about food security.

Although renewed scholarly attention is being given to study of policy implementation (Barrett, 2004; Patashnik, 2008), there remains scant implementation research focussed at the international-domestic policy interface, and the role of international level factors on policy implementation of domestic designs remains inchoate. As noted, our claim is that part of the resilience in the implementation phase of policy in the face of evident disruption mechanisms from the international level is related to policy design. For a policy design to function intertemporally requires the creation of expectations of future policy paths. The credibility of policy designs which commit governments to future choices in implementation, in response to various contingencies, is critical to policy success. If future decisions are expected to coincide with the original policy design (that is, if policy-makers are credible in commitments to stick to the plan), then the policy design is time consistent. If, however, the policy-makers are anticipated to want to renege on the original plan, then the policy design is time inconsistent.

Patashnik (2008) argues that to overcome tensions and difficulties of implementation requires continued political attention from the coalition that enacted the policy initially. He draws on historical institutionalism's aphorism – that policy begets politics – to describe the importance of changes in policy becoming institutionalised and shifting the political incentive structures in the implementation phase. Policy design is one possible source of stability in the face of novel information and ideas mechanisms that all may come into play with implementation. For Patashnik, the agreement of policy is only ever the start; policy is not a one shot game but rather a repeated political game in which the inherent features of political life apply after policy is made. Policy design is an essential feature of that political game: in particular, the extent to which design institutionalises expectations of time consistent policy commitments.

In cases where uncertainty is designed into policy, the diagnosis of a gap between design and execution as evidence of implementation failure is misguided. Instead, the purpose of policy is to set a time-consistent direction for the future policy path, to impose boundary conditions on what is feasible, and provide resources. Indeed, one means of seeking to establish credibility to this end is to devote political and financial resources to the policy. We should not expect a strict match between design and action, yet implemented policy should be recognisable in terms of the antecedent policy as designed.

In some cases, policy design may correctly anticipate uncertainties, in the sense that a known unknown is acknowledged, and policy is able to be adjusted in implementation to accommodate novel information and ideas and thus remain time consistent. Alternatively, closest to a disruption is where new information or ideas are not anticipated and the credibility of commitments to future policy paths and long-term superordinate goals are threatened.

The degree to which policy design involves several policy layers, rather than a single isolated element, is a contributory variable in explaining resilience against disruption caused by new ideas or information. Specifically, the layers of biofuels policy clearly linked in design to the three broad domestic policy goals – energy security, climate change mitigation and rural-urban equity – act to ameliorate the disruptive potential of international level ideas and information mechanisms. Biofuels policies are part of the EU, Brazilian and US policy-makers' means of delivering these broader policy goals, and to challenge biofuels policy design, disruptions from the international would have to disturb either those broader policy goals, or beliefs about biofuels' ability to contribute to their delivery.

Revisiting biofuels policy design

As the book has already described, policy-makers in all three cases have established mandates for blending biofuels in petrol. These have, in the EU and US cases, been set well above current domestic production and consumption levels. The EU mandate, for 10 per cent of fuels to come from renewable sources by 2020, is expected to be fulfilled largely by biofuels. The US mandate, of 36 billion gallons by 2022, could, depending on future fuel consumption, represent 20–25 per cent of liquid transport fuel.

The problem for policy-makers is that whilst the availability and usability of first generation biofuels makes them attractive politically, they have downsides which have become manifest in implementation. As a result, and as Chapter 6 sets out, biofuels policy design has been bifurcated. Policy-makers are trying to contain the downsides of first generation biofuels, through a combination of pre-implementation and post-implementation measures. They are also making long-term commitments, to support the technological developments necessary to bring large quantities of commercially competitive advanced biofuels, which can limit or avoid the downsides of first generation biofuels, to market. This is being done through the mandates, and through financial support. This assistance is both direct, supporting public and private-sector research activities, and indirect, through loan guarantee schemes.

Policy-makers must seek to forge authoritative policy choices, committing relevant state, civil society and importantly, private business actors to particular actions over several years to deliver those policy choices. This, however, presents a policy dilemma: longer-term goals must be clearly enunciated and remain credible, whilst these uncertainties require flexibility and adaptability in policy implementation. Furthermore, biofuels policy implementation has seen potentially disruptive ideas and information enter the policy-making milieu recently, the impact of which on policies is explored below.

Biofuels policy drivers

As the book has elaborated, biofuels policies are motivated by three factors, underpinned by the renewable nature of the product: enhancing energy security, promoting rural development and reducing GHG emissions. All three motivations are essentially domestic in origin. Intentions for biofuels policy have been founded in the domestic arenas of politics in all three cases, even if they are, in the example of reducing GHG emissions, policy responses to a collective, global, challenge. In the US, EU and Brazil, an alliance of interests coalesced at different times around these three drivers to help policy-makers establish biofuels policies. Within these coalitions, however, the distinct characteristics of the three policy drivers resulted in coalition partners having different, even incommensurable, values.

Energy security can be satisfied by any generation of biofuels. Given the rural dimension of biofuels policy the agricultural lobby, especially in the US (as analysed below), supported the establishment of biofuels mandates which, given technological constraints, would necessarily come initially from first generation biofuels. Further, a great many of advanced biofuel feedstocks would also be produced in rural locations. Those concerned with wider environmental, developmental and climate change concerns favoured advanced biofuels, given their greater emissions-reduction potential and potential avoidance of key downsides from first generation biofuels (see Chapter 9). This potential schism within the coalition was accommodated by adopting the bifurcated policy which demarcated intertemporally first generation from advanced biofuels. The demarcation is explicit in the US mandate. In the EU, advanced biofuels count double towards the usage mandate, but otherwise the role foreseen for advanced biofuels is more implicit.

If we revisit briefly the recent history of EU and US biofuels policies, we can see that whilst not perfectly correlated, there is a stronger association between US policy and first generation biofuels, and, relatively speaking, between EU policy and advanced biofuels (at least insofar as these have the potential to deliver higher GHG emissions reductions). As analysed in detail in Chapter 3, through the 1990s, the EU began to position itself as a global climate policy leader. This, in turn, exposed the lack of domestic policy responses. Perceived success with measures to promote renewable electricity led to attention turning to renewable transport fuels, resulting, in 2003, in the Biofuels Directive (EU, 2003, the preamble to which also references energy security), thence the RED in 2009 (EU, 2009a).

As Chapter 4 describes, the first substantive US biofuels policy episode came in 2005, with the passing of the Energy Policy Act (US Senate and House of Representatives, 2005). This included the RFS – a mandated volume of ethanol to be blended into petrol. Interviews revealed this particular aspect of policy was inserted as a direct result of the lobbying from corn and ethanol interests. This ensured continued demand for ethanol, given uncertainties over the status of legislation governing oxygenates in petrol and the role of oxygenates, in particular, MTBE, in terms of damage of the ozone layer and the possible pollution of water sources (McCarthy and Tiemann, 2006). This mandate was developed considerably by the 2007 EISA (US Senate and House of Representatives, 2007), in particular through the much-expanded biofuels mandate in RFS2.

Interviews also revealed the key role played by the Senate in getting RFS2 included in the EISA – the Senate being strongly domestically focused politically and thus concerned with energy security, but also with strong rural/agricultural representation. On the other hand, the commitment to GHG emissions reductions in RFS2 came from the House of Representatives (against Senate opposition). Moreover, a staffer with one of the key committees in one of the Houses of Congress stated that some environmentalists saw this as a bridge to working with agricultural interests (between whom there was a long history of antagonism) – an alliance which therefore helped bind the biofuels coalition.

Thus, as seen in detail in Chapter 5, the surrounding policy discourse indicates EU policy was motivated more by the GHG emissions driver, whilst US policy was driven by energy security and agriculture (specifically 'big ag' interests) – notwithstanding the presence of all three drivers in discourse around each policy. In terms of agriculture's influence on US policy, two features are important. First, the first generation part of the mandate, in effect, ringfences a portion of the mandate for ethanol derived from agricultural feedstocks (in practice, mainly corn). Second, as a means of providing a market outlet for farmers' crops, a mandate specified in terms of an absolute volume of ethanol offers greater certainty than other approaches – such as that of the EU, expressed as a percentage of transport fuel. It is important to recognise that the first generation part of RFS2 is not specifically set aside for ethanol based on the US-produced feedstocks, let alone specifically corn. That said, the only other major ethanol producer at the time was (and still is) Brazil. Its ethanol is derived from sugarcane and is capable of much greater GHG emissions reductions. Brazilian sugarcane ethanol is defined as an 'advanced' biofuel, as a result of which it is located in a separate part of the RFS mandate. An interview with officials from one of the key federal government departments confirmed that the Brazilians were 'astonished' at this favourable treatment.

The EU approach to the specification of the mandate, as a percentage of the fuel mix, raises two distinct issues which contrast with the US 'big ag' approach. First, the volume of biofuels to be blended (and thus the agricultural feedstocks required) will depend in part on total petrol and diesel sales (bearing in mind that, even with the blendwall in the US, the conventional ethanol mandate in RFS2, set as an absolute volume, accommodates more or less exactly the corn-based ethanol resulting from the production of 'big ag'). Second, by not distinguishing between different biofuel types or generations, there is no part of the EU mandate set aside for biofuels derived from EU-produced feedstocks. Thus, in the short to medium term, each first generation feedstock is judged against the same GHG emissions threshold, as set out in Chapter 3, whilst in the future, better-performing advanced biofuels could take the market share of first generation biofuels, if competitively priced – but especially from 2017 and 2018, when the emissions qualification thresholds, defined as part of the EU sustainability criteria, are set to rise significantly.

Interviews in Brussels confirmed that a biofuels policy providing guaranteed markets for farmers' crops would contradict the move to progressive de-coupling of support under the Common Agricultural Policy. This, and the more environmentally oriented focus of EU biofuels policy, is reinforced by the EU approach to the required GHG emissions performance of biofuels. As the figures in Chapter 3 showed, unlike the US approach, EU minima are biofuel- and feedstock-neutral. Moreover, under EU policy, older biofuels processing plants do not receive grandfathering protection.

Implementing uncertainty: The international policy arena and domestic policy design

It is in the implementation of biofuels policy that disruption mechanisms are observed. The expansion of first generation biofuels production, in implementation, has brought with it the growing rev- elation of (actual or potential) downsides. These have resulted in grow- ing opposition to first generation biofuels. Second whilst policy for advanced biofuels allowed time for the development of, at least, second generation biofuels, their continued under-delivery to market is result- ing in greater opposition to biofuels policies generally. There is a clear advanced-biofuel dimension to US policy disruption, as a result of the separate mandate for cellulosic ethanol but, overall, the challenges to biofuels policies tend not to distinguish greatly between first genera- tion and advanced biofuels: all biofuels are being subjected to disruptive pressures.

The policy responses, analysed below, combine design measures (built into the legislation) and implementation measures (notably ongoing monitoring and reporting procedures). Both the EU and US mandates require biofuels to deliver GHG emissions reductions compared with the fossil fuels they replace. The life-cycle emissions calculations of biofuels are, therefore, central to the implementation of policy. This, however, raises a number of issues. First is the complexity of the basic calcula- tion, given that every feedstock/technology pathway combination will deliver a different emissions figure. Second, there are issues over what exactly should go into such calculations (see Chapter 9 for an in-depth analysis of these challenges).

One ex ante policy is the establishment, in both the RED and EISA, of sustainability criteria, as set out in Box 3.1 and Box 4.1, respectively. These are designed to avoid agricultural practices which release GHGs (and, in the EU case, to minimise the possible impact on biodiversity, again reflecting concern with environmental aspects). For biofuels to count towards the EU mandate, two types of cropping potential are excluded. In the US, meanwhile, sustainability criteria are inclusionary in terms of land-use and feedstocks/inputs. With US biofuels (princi- pally ethanol) coming from domestically produced corn, and with the land used for growing it being cultivated already as part of the estab- lished 'national farm', US sustainability criteria have faced no significant disruptions on sustainability-related grounds from anti-biofuels inter- ests, through either ideas or information channels. EU criteria, however, despite their positive intentions, have faced significant disruptions on several grounds, as elaborated on in both Chapters 3 and 9.

This also hints at the transboundary implementation problem the EU faces, given EU biofuels imports are produced along international supply chains.[3] The primary means of delivering biofuels derived from sus- tainably produced feedstocks (as defined by the EU criteria) is through product certification, often undertaken by non-state actors such as Bonsucro and the Roundtable on Sustainable Palm Oil (RSPO). The certification process requires organisations to submit their schemes to the European Commission, who judge the certification standard against the EU criteria. If approved, all biofuels carrying that certification are deemed to conform to EU standards. An important point made to us by a UNICA[4] representative is that the standards must be workable in terms of what producers can deliver, independently of the *process* of certification.

The certification approval process is time-consuming, resulting in significant volumes of biofuels imports currently not certified as sustainable. That said, despite the RED only having been agreed in 2009, certification is making strong progress. In the UK, for example, 31 per cent of biofuels were certified in 2009–2010, rising to 53 per cent in 2010–2011. The estimate for 2013–2014 indicates a figure of 98 per cent.[5]

A further concern related to the sustainability criteria, also addressed pre-implementation in the writing of the RED, is whether they constitute trade barriers under WTO rules. Some consider sustainability criteria intrinsically to represent barriers to free trade.[6] Meanwhile, as analysed in Chapter 2, the Brazilian government's position is to take no position on Brazilian producers' participation in certification schemes, to avoid giving the impression of official validation for those schemes. Instead, they are observing implementation, lest trade concerns arise at this stage of the process and supply chain. The WTO itself, in the Technical Barriers to Trade Agreement (TBTA), provides guidelines over how countries should develop rules in order to reduce the chances of an action being brought subsequently. An important finding from interviews within the European Commission was the extent to which potential WTO concerns were borne in mind when developing sustainability criteria for the EU – in ways we analyse in detail in Chapter 8.

One salient challenge for the implementation of biofuels policy is the accuracy of the GHG emissions calculations. This raises an issue that has been a source of significant disruption: ILUC. As introduced in Chapter 1 and analysed in detail in Chapter 9, ILUC occurs as a result of the overlap between biofuels and agricultural commodity markets. It is particularly vulnerable to disruption because it cannot be observed directly, only modelled. Wide variations in ILUC estimates compound uncertainties over biofuels' GHG emissions performance. Interviews in both Brussels and Washington revealed that it was with the publication of Searchinger *et al.* (2008) that opposition to biofuels on the basis of ILUC effects took off. In particular, environmental NGOs were concerned that, outside of the bounds defined by sustainability criteria (which only cover *direct* land-use changes), biofuels would indirectly impact on emissions and biodiversity. Thus far, in both the US and Brazil, interviewees in policy circles repeatedly expressed considerable doubts about ILUC. One very senior and experienced official interviewed in Brazil argued that even the worst ILUC would not result in emissions estimates rising by more than 10–15 per cent – a figure which is considerably below some of the estimates analysed in Chapter 9.

As an idea, ILUC gained a lot of traction very quickly with anti-biofuels interests. Yet amongst policy-makers, especially in the US and Brazil, the disruption this debate has caused has not affected policy. Indeed, interviews in Washington and Brasilia revealed a similar response. In both countries, policy-makers were continuing to work to improve their understanding of ILUC processes in order to better estimate possible ILUC impacts of their biofuels policies. In both cases, this provided information which challenged the idea that ILUC was a significant problem. Thus policy-making was disrupted, but no change in policy resulted. Indeed, senior officials in Brasilia paid tribute to those who raised ILUC as a potential issue, as it forced them to analyse the issue more carefully. This process, they believed, confirmed the excellent environmental performance of Brazilian sugarcane ethanol.

In contrast to the US and Brazil, there has been an EU response to ILUC debate. As discussed in detail in Chapter 3, reform proposals were published in 2012 which would introduce a series of flat-rate ILUC factors into biofuels' emissions calculations. That said, agreement is not expected until late 2014 or early 2015 at the very earliest. Moreover, even if ILUC factors are added into emissions calculations under the EU sustainability criteria, with the significant increases in GHG emissions-reduction thresholds already scheduled for 2017 and 2018, it is not clear exactly how much of an *additional* impact the ILUC factors would have, in practice, on feedstock choices.

From ex ante prevention to ex post monitoring

Sustainability criteria have been designed with the intention of possibly irreversible damage to land and biodiversity. They have also been designed to avoid violating WTO trade rules. First generation biofuels, however, also have other side effects, the magnitude of which is not known ex ante, even if their potential occurrence is. Note that ILUC is different in that even though its existence and possible magnitude are still being understood, ultimately it could be embedded in the legislation in some way, as we see the EU proposing, to avoid damaging land-use changes happening.

The policy responses for these other concerns are for monitoring to occur post-implementation. Potential problems arising from the production of biofuels and feedstocks that are being addressed in this way include social sustainability and labour conditions, and the 'food v. fuel' debate and the possible impact of biofuels on food prices (for more on which, see Chapter 9). This arises through both the use of agricultural feedstocks for biofuels instead of food and animal feed, and through land-use change effects (direct and indirect), occurring as a result of price transmission from biofuels to agricultural and food markets.

Social sustainability and labour standards were discussed in the EU when the RED was being drafted, with the European Parliament strongly supporting their inclusion (Daugbjerg and Swinbank, forthcoming). Within the Commission, however, there was concern that including social criteria would cross some countries' 'red lines', risking an action being brought in the WTO that could threaten the whole biofuels policy (Lydgate, 2012b, refers to this influence on policy design as 'regulatory chill'). That said, the debate continues as to whether such criteria could be included without violating WTO rules, not only within the EU but also in International Standards Organisation debates on potential biofuels standards.[7]

Other than in this specific context, the question of social standards in biofuels policies did not arise in our fieldwork. Domestically, there is a strong social dimension to Brazil's biodiesel policy, as analysed in Chapter 2. Indeed, Brazil has been implementing improved standards in the sugarcane sector for some years (see, for example, Schaffel and Lèbre La Rovere, 2010; Schaffel *et al.*, 2012; Labruto, 2014). Moreover, as discussed in Chapter 8, the certification bodies that the EU is working with to implement its sustainability criteria have their own established criteria which, unlike the EU version, do include concerns such as labour, social and human rights standards. That said, international trade rules (see Chapter 8) have different implications regarding what policy-makers and private sector companies can do.

Externally, however, the challenge – as one interviewee in the European Commission put it – is working out how exactly one can legislate for this. Partly this comes back to the WTO issue. Whilst Brazil, for example, has been working on improving domestic labour conditions – and therefore may conceivably have attained standards that could withstand international scrutiny – other countries may not be in this situation. Moreover, the question of labour and social standards is not unique to biofuels – so are those who work to produce feedstocks for biofuels going to be singled out for special protection? In this regard, discussion over labour standards and whether it is even feasible to incorporate them into WTO-compatible policies and agreements has strong parallels with the issue of Processing and Production Methods, analysed in detail in Chapter 8.

One paper that looks specifically at social issues and biofuels sustainability criteria is that by Charnovitz *et al.* (2008). They argue (on page 2) that one unique aspect of social concern arising from biofuels is that of food security and the impact of biofuels on food prices. But, as we analyse in detail in Chapter 9, the impacts of biofuels on food prices are, at best, ambiguous. Thus would social criteria lead to a given batch of biofuel being accepted in an importing country only once a detailed analysis of the impact on food prices of the feedstocks used to produce that biofuel been conducted? Furthermore, they argue that another reason to single out biofuels is the impact government policies have had on promoting the production of feedstocks, where the social/labour concerns arise. This, however, assumes that biofuels represent the only market where such interventions take place. In short, the two arguments given by Charnovitz *et al.* for singling out biofuels for special social/labour treatment are, at best, debatable.

This is not intended to downplay the normative value of social concerns, but rather illustrate the challenges faced in the possible design and implementation of social/labour standards. Such arguments for standards are, without question, made in good faith, but as the above discussion shows, the assumptions on which policy could be based are controversial – in ways we analyse in detail in Chapter 9. Even voluntary reporting, as adopted by the EU, can help shine a light on bad practice. If this is combined with commercial pressures (for examples in the way companies are increasingly seeking to ensure beef, leather and so on do not come from animals reared and grazed on deforested land in Brazil's Amazon), then this approach is not without impact.

Perhaps the most significant contribution to this debate in recent years is Nuffield Council on Bioethics (2011). This extensive report identifies six ethical principles which they argue should guide biofuels production decisions:

- Biofuels development should not be at the expense of people's essential rights (including access to sufficient food and water, health rights, work rights and land entitlements);
- Biofuels should be environmentally sustainable;

- Biofuels should contribute to a net reduction of total greenhouse gas emissions and not exacerbate global climate change;
- Biofuels should develop in accordance with trade principles that are fair and recognise the rights of people to just reward (including labour rights and intellectual property rights);
- Costs and benefits of biofuels should be distributed in an equitable way;
- If these first five principles are respected and if biofuels can play a crucial role in mitigating dangerous climate change, then depending on additional key considerations, there is a duty to develop such biofuels. These additional key considerations are: absolute cost; alternative energy sources; opportunity costs; the existing degree of uncertainty; irreversibility; degree of participation; and the overarching notion of proportionate governance.

Even here, however, the uncertainties remain over how work rights, labour rights and so on, can best be implemented in policy. One issue faced with any policy, let alone that of the EU, is that the sustainability criteria cannot guarantee absolutely compliance with work rights and all of these other criteria. Moreover, two-yearly reporting may reveal problems, but it might be too late by then to address specific violations of these principles.

The potential impact of biofuels on food prices has been under particular scrutiny since 2008. In that year, commodity and food price levels and volatility rose sharply, as did biofuels production. This led to development NGOs challenging biofuels, given the effects on food prices, supplies and food security in developing countries. Subsequently, however, the fall in food prices has not stopped development NGOs from continuing to campaign against biofuels, focusing increasingly on the local social impacts of land-use changes, as well as food security issues. Campaigns against biofuels and their perceived impact on food prices also emerged in developed countries, notably in the US where they were led by the Grocery Manufacturers Association (GMA). Whilst they were less vocal in 2009, when US biofuels production continued to rise but commodity prices fell sharply, one of our interviewees made the point that the coalition, established initially to campaign over food prices, has remained active but shifted its attention more to other biofuels-related issues, such as the debate over introducing the E15 petrol blend and the (now-ended) biofuel blenders' tax credit (as discussed in Chapter 4).

Again, however, despite the challenges these ideas have presented to policy-makers, they have resulted in little policy change thus far. As with land-use change, prices, demand and supply in biofuels, commodity and food markets are influenced by many factors, making determining causality rather than mere correlation very difficult – an issue we explore in detail in Chapter 9. Whilst the EU reporting process includes monitoring the possible food price impacts of biofuels, policy-makers have been reticent to accept a causal link. With debate still at the level of ideas and with information over the magnitude of links still strongly contested, there has so far been no policy shift.

Ultimately, whilst there are multiple domestic–international linkages, our interviews revealed very little clear evidence of ideas at the international level shaping or re-shaping domestic policies. The one notable exception has been the influence of WTO trade rules on the ex ante formulation of EU sustainability criteria. This lack of impact is despite the idea of sustainability, with its exhaustive, totalising logic, being difficult to manage intertemporally, as both the EU and US have done with their first generation and advanced biofuels policies. Although the idea of sustainability is so strongly contested that information flows about the environmental consequences of biofuels production and consumption can be disruptive and potentially undermine the implementation of an institutionalised, shared view of biofuels policy, this has not happened. This is so even as concerns over food security have gained currency, with linkages made to other issues and values such as energy, the environment, economic development, agriculture and trade. In Chapter 9, we develop some of these ideas further, notably regarding how media representation can influence public opinion on biofuels. That said, whilst we argue there that better engagement with consumers can improve support for biofuels, the principal theme of this analysis – that there has yet to be fundamental change to policy as a result of disruption through any of these various channels – remains intact.

The interaction of market uncertainty, the implementation gap and the international policy process

Our interview data show that reformers' attention has waned in the US (with key individuals in the Senate gone or less influential), and in the US and EU, that the initial spirit of the reforms that held the reform coalition together has been contested in implementation, as outlined above. Notably, both cases reveal that the international-level discourse of sustainability, far from helping to integrate various policy objectives in the implementation phase in a common concern, can, first, add complexity to initial policy design and increase the number of objectives, interests and players in policy implementation, and, second, be used as a key resource by institutionally embedded actors in existing policy legacies to protect their autonomy against the ambitions for a coordinated implementation of the initial policy design.

The feedback processes from policy adoption to subsequent policy changes are not clear, and, moreover, not obviously positive feedbacks but rather some uncertain mixture of positive and negative. There is evidence of policy layering in design, for example, as seen in Chapter 6. But policy layering in design needs to be complemented by an analysis of its consequences in implementation. In particular, the interaction between different legacies and new interests can be observed in the lack of institutionalisation and coordination between the two parts of the design in the implementation phase. The biofuels cases are suggestive of the particular difficulties of joining up government in the international governance of sustainability. The two part policy designs enacted in the US and EU immediately introduced a number of bureaucratic players in implementation.

In the US, the Departments of Agriculture, Energy and State, plus the EPA, were prominent. Within the European Commission, several Directorates General (DGs) were involved, notably Energy, Environment, Climate Action, Agriculture and Transport. To these can be added a wide range of other stakeholders and interested actors that have been part of the implementation of policy. Furthermore, as noted, the emergence of new international bodies such as GBEP provides a forum for discussion between the major players in a more constructive and progressive manner than traditional international intergovernmental organisations.

Lack of coherence in policy design, as already suggested, is positively related to lack of coordination in policy implementation. The formal institutions of US and EU systems are relatively autonomous and non-hierarchical, with multiple access points for the ideas and information mechanisms to exert leverage. In addition, stimulating a large-scale biofuels market is a new policy agenda. As a result, the old bureaucratic structures have, in recent years, undergone significant overhaul. This has led to the introduction of a new unit in the EU's Directorate-General for Agriculture (interestingly, under 'rural development'); the single DG for transport and energy has been split into two (Energy and Mobility); and DG-Environment has seen part of its activities moved to a new DG for Climate Action. Meanwhile, between our two visits to the US State Department in 2010 and 2011, a new Bureau for Energy Resources had been created, wherein biofuels would be located.[8]

This institutional *milieu*, itself unstable, facilitates a rapidly moving and potentially disruptive set of ideas and information flowing into policy, given the different interests involved and the porous nature of policy-making institutions. The key dimension on which to evaluate the impact of ideas on policy-making is integrative-disintegrative, that is the degree to which ideas help mould existing policy legacies with new policy interests in biofuels into a shared sense of a coordinated policy. This is not the same as agreeing, or that everyone has the same values or interest, but rather there is a shared view of the policy frame within which policy has been designed and implemented. As yet, the ideas and information flowing into policy have not disrupted this shared view of policy.

Policy layering is an important feature of biofuels policies in the EU and US, but so too is the nesting of biofuels policies in wider agendas. In the EU, the RED was negotiated through 2008 as part of a climate and energy package, as analysed in Chapter 3. This sought to deliver a strengthened EU Emissions Trading Scheme (ETS); an agreement for member states to share efforts to reduce emissions in sectors not covered by the ETS; binding national targets for renewable energy (including transport fuel); and plans for development of carbon capture and storage technologies.

In the US, the EISA is itself much broader than just biofuels. Further, parts of Title II of the EISA, 'Energy Security Through Increased Production of Biofuels', directly amend the Clean Air Act. The 2008 Farm Bill had several measures aimed, primarily, at promoting advanced biofuels development, given concerns over potential problems from expanding corn use for ethanol (Chite 2012: 45). We see this also in the new 2014 Farm Bill.[9] Furthermore, biofuels policies include a variety of fiscal measures, including research support, tax breaks, etc. The current state of the debate in the US suggests broader fiscal concerns, as outlined in Chite (2012), may pose as big a threat to the future trajectory of biofuels policies as food prices or sustainability.

Thus, to challenge biofuels policies, disruptions mechanisms would have to challenge either these broader policy agendas, or biofuels' ability to contribute to their delivery. This brings us back to the three drivers of biofuels policy and issues introduced in earlier chapters, notably the ability of biofuels to be defended against an attack on one of the drivers, by reference to one or both of the others. The counter-intuitive result is that, as complex as such a policy is, it is precisely the layering of policy and the nesting of biofuels within broader debates that has helped protect the original, domestically motivated design against novel and unanticipated domestic and international pressures explored in this and earlier chapters. Mondou *et al.* (2014), in their detailed analysis of the resilience of US policy, point to findings that, they suggest and we agree, apply also to EU policy.

They point to modest changes being made to policy in response to negative policy images, whilst simultaneously responding more pro-actively to other negative images. Examples here from the US would be the ethanol lobby not campaigning against the expiry of the tax credit, in the face of the fiscal cliff, whilst putting out an aggressive counter campaign against the GMA-led campaign in 2008, pointing out (for example) how small a share of the total cost of a box of Corn Flakes was represented by the corn (5 cents). Even the threat of biofuels' contribution to energy security being undermined by the recent shale gas boom is being countered, partly, through the support of the US Navy for advanced biofuels – against opposition from the House of Representatives (Mondou *et al.*, 2014: 175). This study thus illustrates how, with a degree of flexibility over some elements of policy design, policies with essentially domestic origins can, in their implementation, display the potential to accommodate and deflect new threats to their design integrity.

Conclusions

Brazil, the EU and US biofuels policy have been designed, and enacted, principally as domestic political actions. This chapter has charted the increased salience in domestic political agenda of the international dimensions of biofuels production. Domestic policies have been driven by concerns over energy security, GHG emissions reductions and rural development. In implementation, however, the multiple dimensions of GHG emissions calculations for biofuels have become the core feature of politics in both jurisdictions. In addition, there are linked concerns over the potential trade policy impacts of the policy response to sustainability concerns. The biofuels policy agenda in implementation has also been upset by the highlighting of undesirable impacts of biofuels, for example on land-use change, and the food v. fuel dilemma.

These concerns and their advocates are international in orientation. Even though the initial focus of biofuels policies in all three cases was domestic, the crucible of biofuels policy agenda has shifted quickly to the international level. Thus far, however, there has been no substantive policy response to these potentially disruptive influences. In all three cases, whilst policy-makers are aware of such challenges to biofuels policies, the design of these policies contains different layers connected to broader policy goals and strongly embedded values, of security, rurality and the environment. Potentially disruptive ideas from the international level triggering debates over biofuels policies have, so far, failed to challenge these broader values, and the different layers in the policy design to which they are linked. This has been reinforced by the contested nature of information which underpins these values, and the ideas which have threatened them but, thus far, have done so without policy-makers responding by changing the policy. But by what channels can international threats to policy be manifest? This is a distinct question from the nature of those threats – and one we analyse next.

8
External Dimensions to Biofuels Policies[1]

Introduction

One of the consistent themes to have emerged from our interviews for this project was that biofuels policies, initially, were established as domestic policies. Over time, however, as biofuels policies have stimulated growth in biofuels markets, external dimensions have come increasingly to the fore of the policy debate, as analysed in Chapter 6 and, in particular, Chapter 7. As Part I demonstrated, as a domestic construct only EU policy has had a significant external dimension to it, given the need for imports to enable the EU to deliver on its policy mandates. Other policies did not preclude trade, but nor did they include imports as a core dimension of delivery on the (domestic) policy. That said, interviews revealed that Brazil's government has sought to promote the development of an international market in biofuels, within which biofuels can be traded freely, just like other commodities.

One aspect of the external dimension to biofuels policies has simply been an increase in biofuels trade – and we look at certain (policy-related) aspects of biofuels trade below. Beyond this, however, a number of significant policy issues have taken up more and more of policy-makers' attention. In this chapter we focus in particular on two of these external features. The first is aligned broadly with Chapter 6 and, in particular, policy design; the second is aligned more with Chapter 7 and implementation challenges.

The first external factor we consider is the World Trade Organisation (WTO) and the rules under which international trade is conducted. Specifically, domestic policies must not distort trade. This applies to policies located 'behind the border' (i.e., domestic policies which are not directly trade-related, but which impact on the conditions of trade), as well as those domestic policies which operate 'at the border', such as import tariffs. This has raised an interesting challenge for policy-makers. More often than not, markets and their associated policies which come under scrutiny are already in existence. As a result, a domestic policy may need to change as trade policy rules evolve. Alternatively, a country's domestic policy may come under threat if it is deemed by one or more other WTO members to have damaged their legitimate economic interests under WTO rules, sufficient to warrant taking that country through the WTO's Dispute Settlement Procedure (DSP).

In the case of biofuels policies, however, in the US but in particular in the EU, the challenge is to ensure entirely new policies and policy instruments are WTO-compatible. This affords policy-makers the opportunity to build-in features reflecting WTO rules. Although we shall see later, examples of how this was done, it is by no means straightforward. A central problem is that even if certain types of policy instrument are actionable under WTO rules, an action will only be brought if one or more WTO members deem it worth doing. Thus some policies may never be tested (or may never yet have been tested) by the DSP even if, on paper, they appear to breach WTO rules. We shall see later one such case with biofuels, with the application by the US of a duty on imports from Brazil. In that instance, the policy instrument used was unquestionably in violation of WTO rules – the question is, why was an action not brought by Brazil in the WTO? This problem of establishing definitively the WTO compliance (or not) of policy instruments is even greater if a new policy brings with it novel policy instruments, entirely untested in the WTO context. Again, we shall see examples of this later, in particular with the establishment of the EU sustainability criteria.

It is these sustainability criteria that bring us to the second significant external factor presenting a challenge to policy-makers. In order that the biofuels blended into fuel in the EU (or the US) conform to domestic sustainability criteria, a robust system of monitoring and traceability must accompany the sustainability criteria. This must be capable of ensuring that the production conditions of different feedstocks, produced by individual farmers thousands of miles away, possibly in remote parts of very large countries, conform with the sustainability criteria laid down by EU and US policy-makers. In this chapter, we shall focus on the implementation of the EU criteria, as the nature of EU policy, with its much greater emphasis on imports than US policy, generates greater concern in this regard. This is reinforced, first, by the fact that EU policy has the delivery of climate change mitigation and positive environmental benefits at its core; and, second, because with its greater use of biodiesel, EU policy connects more directly to palm oil and questions of deforestation in countries such as Indonesia and Malaysia.

This creates an implementation problem insofar as the government, the EU, is unable directly to control and manage the entire supply chain. Instead, it must rely on a variety of partners who work with it to ensure the terms of the policy are respected. These will include bodies who undertake the monitoring of the production conditions of the agricultural feedstocks and certify that the sustainability criteria are being respected; companies to audit the certification process; and (multinational) companies to purchase the feedstocks, convert them to biofuel, buy the biofuel and, finally, blend them into transport fuel. Although the EU has rules in place which seek to ensure all parties deliver what is required to enact the policy as desired, the EU must also rely on the efficiency of each party, and the efficacy of the monitoring process at various stages along the lengthy supply chain.

More formally, this challenge is seen not as a challenge of government, but of governance. The latter term has various meanings in different contexts. In this context, we use the term to describe a mode of policy management which requires the 'government' to engage directly with a variety of public *and private* sector actors to ensure policy delivery. This is sometimes described also as a shift from hierarchy to network – concepts introduced in Chapter 6. In the former, the government would have a direct role in influencing what other parties did. In a network, the relationships are looser and less direct. It is based more on cooperation, notwithstanding the contracts that may be put in place, the fiscal incentives the government has at its disposal, and the policy steer it seeks to give in this way.

This shift, from hierarchy to network, is itself seen as a manifestation of the globalisation of economic activity, such that even if a government retains internal sovereignty, within its national borders, the expansion of international trade and especially the internationalisation of supply chains weakens a government's external sovereignty, its ability to control economic activities and exchanges beyond its borders. The more extensive those global chains, the greater the share of total economic activity which takes place beyond the direct control of a given government.[2]

The rest of this chapter considers, first, the range of WTO-related policy issues that have arisen in the design (and implementation) of biofuels policies. We then explore in more detail issues of globalisation, governance and the transboundary implementation of domestic policy.

International trade – Policy-making in the shadow of the WTO

The WTO has no direct role to play in domestic policy-making. Indeed, as it describes itself, the WTO is essentially the table around which the many countries that are members of the WTO sit to make policy. It facilitates this process, and it manages the process when a trade dispute arises between two or more members. The influence of the WTO, in the main, is indirect – it casts a shadow over national policy-makers, who should take account of WTO policy, trade rules and the obligations on members towards each other, when designing and implementing domestic policies (what Lydgate, 2012b, calls 'regulatory chill').

We start by outlining the range of WTO-related dimensions of policy that might affect biofuels policy-making. In this regard, one particular feature of first generation biofuels is especially significant. In the WTO, trade in agricultural goods is subject to different rules, compared with industrial goods. As a result, many WTO member countries, including the EU and US, have extensive and complex systems of policy intervention in the production and trade of agricultural commodities – including those which are also utilised as feedstocks for biofuels. This has an important consequence for the consideration of biofuel policies in the WTO context.[3] We see the consequences of this in both the present section, but also later, when we look at trade in biofuels products directly.

Table 8.1 summarises the main biofuels policy instruments which raise possible concerns in the WTO. This table also includes reference to the Doha Round Modalities. These are policy issues which remain on the table, gathering dust, given the failure to deliver a comprehensive agreement on the WTO Doha Round of trade liberalisation talks despite more than a decade of negotiations. A point we return to in Chapter 9, in particular, is that biofuels policies and the debates they generate are ongoing, and this is but one more manifestation of this. One theme which is prominent in debates around biofuels policies and the WTO is whether or not biofuels can, or should, be treated differently to other traded goods. The dominant view is that biofuels should not be treated differently, in and of themselves. As Table 8.1 shows, however, there is discussion ongoing in the current Doha Round of trade negotiations about whether goods classified as 'environmental goods' can or should be treated differently from other traded commodities. This discussion remains on the table, as yet far from resolution (one interviewee in Brazil

Table 8.1   Biofuels policies and WTO trade rules

WTO provisions	Issue raised	Biofuel policy
GATT articles I, III, XX	Non-discrimination National treatment	Blending mandates Biomass tariffs Biofuels tariffs
TBT agreement	Trade implications	Standards
SCM agreement	Financial contribution Conferral of benefit Specific subsidy Adverse effects Nullification or impairment of benefit to trading partners Serious prejudice to another country	Capital grants Low-cost loans Tax treatment Ethanol subsidies Tax credits Research and development grants
AoA (annexes 1 and 2)	Domestic support AMS limit Domestic support Green Box eligibility	Biomass production subsidies (direct and indirect)
Doha round modalities	Environmental goods definition Tariff cuts in industrial goods Agricultural tariff cuts Agricultural subsidy cuts Changed definition of Green Box	Improved market access Biodiesel tariffs Ethanol tariffs Domestic support reductions

Note: All acronyms, terms and policies are explained in the main text.
Source: Josling *et al.*, 2010: 30.

was clear that from their government's perspective, there will not be agreement on the meaning of environmental goods in the WTO unless the list includes biofuels). The significance of the issue of how to treat biofuels as a traded commodity will be seen in the following analysis.

When the WTO was created in 1995, it absorbed the GATT administration. It also updated and subsumed the agreement on tariffs and trade. That said the original agreement, 'GATT 1947' formed the basis of 'GATT 1994', the latter also incorporating a series of additional documents.[4] The essence of the GATT as an Agreement thus remained untouched. What did change with the Uruguay Round of negotiations, which began in 1986, concluded in December 1992 and saw the agreements signed in Marrakech in 1994, was that it consisted of negotiating, in parallel, a number of complementary agreements. The principle underpinning this approach is the 'Single Undertaking' or, informally, 'nothing is agreed until everything is agreed'. These negotiations included agreement on the Technical Barriers to Trade Agreement (TBTA) Subsidies and Countervailing Measures Agreement (SCM) and the Agreement on Agriculture (AoA), alongside the GATT – all of which now sit underneath the WTO umbrella.

GATT provisions

Article I of the GATT sets the most-favoured nation (MFN) principle. Simply put, this requires each country to offer to all partners trade arrangements equal to the partner facing the lowest trade barriers – the country's most-favoured partner. This applies at the level of each individual product, with a requirement that MFN applies to 'like products' – products which are, at some essential level, the same as each other. This is a key concept to which we shall return on several occasions. So what are like products? In hearing an appeal as part of one particular case,[5] the WTOs Appellate Body (AB) drew on an earlier GATT Working Party Report to produce a list of four criteria that, whilst 'neither treaty mandated nor a closed list of criteria', help establish product likeness (Condon, 2009: 906):

- Sharing physical properties, nature or quality
- Serving the same or similar end-uses
- Whether consumers perceive or treat the products as serving the same or similar end uses
- Sharing the same international tariff classification

The first three factors, in one way or another, relate to demand and use of a product, not the production or supply side. The significance of this will be seen clearly later, when we discuss the concept of, and problems associated with, Processing and Production Methods (PPMs) in WTO law. The fourth factor raises another problem – there is no simple unified codification of ethanol and biodiesel in the Harmonised Commodity Description and Coding System of international tariff classification. Some of the distinctions are based on different feedstocks/inputs, others are based on end-use. This also complicates establishing just how much ethanol and biodiesel are traded each year for transport fuel purposes.[6] Ultimately, any WTO Panel or AB would have to determine likeness case by case, reducing to pure speculation any general ex ante assertion about biofuels produced in different ways being like or unlike each other.[7]

MFN is central to the WTO, but there are two notable exceptions provided for. First, Article XXIV allows for preferential trading agreements (PTAs), the most notable example of which is the EU, whereby members remove trade barriers between themselves, but not with non-members. The second main exception to MFN is set down in Article XXXVI. Developed countries can offer developing countries trade concessions without requiring reciprocal reductions to be made by the latter, as is implied by the MFN principle. MFN and reciprocity are, however, the fundamental building-blocks of the GATT, the WTO and the multilateral trading system. They relate to the general principles by which trade policy instruments are utilised, but not the instruments themselves. For this, we move on to Article III.

Article III of the GATT relates to 'National Treatment on Internal Taxation and Regulation'. This also utilises the concept of like products. In Article I, that concept referred to the treatment of like products from different third countries. In Article III, it applies in relation to goods produced domestically and goods imported into a country. The essence of Article III is that domestic policies cannot discriminate between domestically produced and imported like products. This relates both to fiscal instruments ('internal taxation') and regulatory instruments. That said, non-discrimination does not require the same policy instruments are applied to domestic and imported goods – just that the effects are equivalent. Note also that this does not prevent countries imposing import tariffs on biofuels, so long as the tariffs applied, again subject to certain exceptions, do not exceed the levels laid down in the country's Schedule of negotiated and agreed import tariffs. In other words, there can be tariffs 'at the border' which, by definition, discriminate between imported and domestic goods, but there cannot be discrimination between these goods as a result of policies 'inside the border', once the imports have entered a country.

The references offered through this chapter to provide further, detailed, reading include several written by legal scholars. Necessarily, the legal approach is very different to that of policy scholars such as ourselves. Thus we now turn to a question that, as indicated later, is of more concern to legal scholars than policy scholars. Specifically, since biofuels are, in principle, subject to Article III might they, in turn, be eligible for a General Exception, as laid down in Article XX? Might there be a way for governments to restrict imports of biofuels that is WTO-compatible? Article XX lists ten exceptions which include, for example, the right to restrict imports of products on the basis of protecting public morals, policies relating to the trade of gold and silver, and to protect patents (for example, by tackling trade in counterfeit goods) and so on. There are two particular exceptions which are routinely examined in the academic literature, to explore their possible relevance to trade in biofuels: Article XXb, which refers to measures 'necessary to protect human, animal or plant life or health', and Article XXg, which refers to measures 'relating to the conservation of exhaustible natural resources if such measures are made effective in conjunction with restrictions on domestic production or consumption'.

It is important to note that, in application, Article XXb has generally been interpreted as relating to domestic measures, with Article XXg applied to trade-related issues. In the context of biofuels and the WTO, first generation biofuels are derived from agricultural commodities. These are dealt with in particular in the AoA, discussed below. The prevailing view is summed by Condon (2009: 905) when he argues that the most likely outcome of this debate is that 'Article XX is not available to justify violation of the Agreement on Agriculture'.

We look in detail later at the debate surrounding the EU sustainability criteria and their possible breaching of WTO trade rules. At this stage we focus on one particular legal dimension. The foregoing discussion proceeds on the basis that regardless of whether biofuels are produced in accordance with sustainability criteria or not, they will be 'like' each other and therefore subject to Article III. Thus the follow-up explores whether it might be possible to treat the two types of biofuels differently, via a General Exception from Article III, utilising Article XX. There is an alternative route, in theory at least, to permitting the differential treatment of biofuels produced in accordance with sustainability criteria or not: can they be seen as 'unlike' and therefore not subject to Article III in the first place? This, in turn, utilises the concept of processing and production methods (PPMs), to see if they create sufficient unlikeness in biofuels produced in accordance with sustainability criteria or not.

As appealing as this might be, theoretically, WTO rules and precedent create fundamental problems. First, sustainable biofuels are not physically different from non-sustainable biofuels. Second – and arguably even more significantly – there is no difference between the processing and the production methods of the *biofuels*, regardless of whether or not there are differences in the processing and the production methods of the *feedstocks*. Third, both do exactly the same job as a transport fuel. These 'unincorporated' PPMs (see, for example, Switzer, 2007: 38) are also known as non-product-related PPMs (nprPPMs). As the

WTO explains on its website,[8] many WTO members argue that nprPPMs should not permit discrimination under WTO law. Another argument, made in the form of a specific example, was made to us in a number of interviews. If PPMs, especially nprPPMs, could be used as a basis for trade discrimination under WTO law, countries could then legitimately block the import of all sorts of goods from countries such as China, where emissions from (for example) old-technology coal-fired power stations were very high. And that was not going to be agreed! And so this brings us back to the distinction between policy and legal scholars: as interesting as these debates over legal interpretations are, in practice biofuels are being treated, and should be treated, as any other traded good – and thus such discussions as those outlined above have academic value but, in practical terms, the issue is rendered moot.[9]

TBT agreement

The Technical Barriers to Trade Agreement (TBTA) covers product standards and labelling. A crucial distinction is made between a technical regulation (which is defined as being mandatory), and a standard (which is defined as being non-mandatory). The TBTA is clear that such measures must respect the fundamental precepts of the GATT: non-discrimination, national treatment and so on. Table 8.1 identifies the TBTA with standards, insofar as biofuels are concerned. The particular example we explore in detail below is the EU sustainability criteria.

It was mentioned earlier that a feature of biofuels policies, because they are so new, is that they can be designed explicitly with WTO rules in mind. Moreover the TBTA, rather conveniently, provides advice in its Annex 3 about how to design such measures as fall under its purview. We return to these provisions later. Annex 3 also helps inform the debate around product likeness, which was shown earlier to be of fundamental importance in the implementation of WTO Agreements. Specifically, Paragraph I requires that, '[w]herever appropriate, the standardizing body shall specify standards based on product requirements in terms of performance rather than design or descriptive characteristics'. As observed previously, product likeness is based primarily on demand-side features – which links back also to the analysis of PPMs. When we analyse the EU sustainability criteria later, the practical significance of this will become clear.

SCM agreement

As with the TBTA, the SCM is eponymously named, dealing as it does with subsidies and countervailing measures.[10] It seeks to limit the use by governments of subsidies which confer a benefit on the recipient and thus distort trade (subject only to the exception laid down in the AoA for specified products). Moreover, by way of remedy, it has the option not only of trying to get the subsidising country to remove the subsidy, but also of allowing countries whose domestic economic interests have been damaged by the subsidy to impose countervailing duties. These are duties on imports from the subsidising country and are intended as an alternative mechanism for pressuring the subsidiser to stop.

WTO rules identify two types of subsidy. Some are prohibited outright, such as those linked to exported volumes, or which require the use of domestic rather than imported goods. Both of these are fundamentally trade-distorting and, if an action is brought against one of these, they are dealt with via an accelerated DSP. Actionable subsidies can also be the subject of a dispute but, because they are not subsidies which inevitably and intrinsically distort trade, the plaintiff must demonstrate the economic damage the subsidy has caused. Table 8.1 identifies a range of possible measures, components of countries' biofuels policies, which could fall foul of the SCM Agreement.

Agreement on agriculture

Finally in this section, there is the AoA.[11] This enters the biofuels debate because many biofuels, but especially first generation biofuels, are derived from agricultural feedstocks (as discussed in Chapter 1). As indicated earlier, agriculture is subject to the general rules of trade, except insofar as the WTO agreements determine otherwise, via the AoA. Whilst agriculture was always a part of the GATT process,[12] it was only with the Uruguay Round that agriculture was subject to comprehensive negotiations over measures to liberalise trade. Because this was the first time that agriculture had been addressed in this detail in a multilateral forum, the negotiations were extremely challenging. This was where the significance of the Single Undertaking came into its own, because failure to agree on agriculture would have resulted in the total failure of the Uruguay Round.

At the core of the AoA were agreements on three distinct types of agricultural policy instrument – those that (directly or indirectly) subsidised exports, those that represented restrictions on imports (or 'market access') and domestic (internal, behind the border) support instruments. The agreement under each heading is as follows. Developed (developing) countries had to reduce the volume of subsidised exports by 21 per cent (14 per cent), and spending on export subsidies by 36 per cent (24 per cent). To improve market access, tariffs had to be cut by 36 per cent (24 per cent), with no individual tariff cut by less than 15 per cent (5 per cent).

The story for domestic support is slightly more complicated. On the basis of a partially adopted traffic-light system, policies deemed trade-distorting were classified as Amber Box policies. These were aggregated together, to create an Aggregate Measure of Support (AMS). These policies were then to be cut by 20 per cent (13–14 per cent for developing countries), compared with a 1986–1988 baseline. Policies deemed as having no or minimal distorting effects on trade or production were classified as Green Box. Excluded from the AMS, these policies were not subject to any reduction. Returning to Table 8.1, the key point is whether biofuel feedstock-related support transfers were placed in the Amber or Green Box. If the former the AMS could not, in total, exceed the specified ceiling.[13]

Table 8.1 refers to Annexes 1 and 2 of the AoA. Annex 1 sets out explicitly all of the products subject to the AoA, and is based on the Harmonised System (HS) classification. This includes all agricultural commodities. Annex 2 then sets out the conditions by which expenditures are exempt from the AMS calculation and subsequent cut. This includes Green Box policies, and a range of other measures. One of these (in paragraph 12) is payments under environmental programmes. Whilst biofuels might, themselves, be considered environmental goods (subject to all of the qualifications highlighted throughout this book), it is unlikely that this would apply to growing feedstocks for biofuels. Specifically, paragraph two states that 'The amount of payment shall be limited to the extra costs or loss of income involved in complying with the government programme'. The government programme in question, however, refers to biofuels, not feedstocks per se. Moreover, mandate obligations are placed on the fuel companies, not the farmers growing the feedstocks.

The key point from this analysis of the Green Box is that it does not provide a shelter for support directed at biofuels feedstocks, just because biofuels *may* potentially deliver environmental gains through reduced GHG emissions. Another point to note is that whilst the SCM addresses subsidies and their prohibition, Article 3 of the SCM states explicitly that an exception to the general SCM rules exists for agriculture-related subsidies provided for in the AoA. Even so, this is subject to the limits set out above. We thus reach our key conclusion from the foregoing discussion – that extensive debate thus far on the place of biofuels in the WTO confirms that biofuels are subject to the same rules as any other traded commodity.

Biofuels and the WTO – A study of EU sustainability criteria

Table 8.1 and the subsequent analysis identify several agreements under the WTO which bear on biofuels. That said, the extent to which these concerns manifest themselves in practice is unclear. Even if a policy instrument is actionable under the WTO, it still requires another member to bring an action against it. The reality is that biofuels policies, and the policy instruments of which they are composed, have not been tested in the WTO. As a result, much of the research undertaken has been speculative in nature. This has focused principally on two key issues: whether certain biofuels policy instruments might violate one or other elements of the WTO Agreements outlined above, and whether, within the WTO Agreements, one of the provisions made for exceptions to the general rules might apply to biofuels policies – as analysed in detail earlier.[14]

'The main problem the EU faces with respect to the WTO is over its environmental sustainability criteria for biofuels.' (Swinbank, 2009b: 501). We thus focus in particular in this chapter on this policy question – and the EU sustainability criteria introduced in Chapter 3. In so doing we draw, in particular, on Ackrill and Kay (2011b), itself written as a follow-up to Swinbank (2009b).[15] Swinbank (2009b: 501) concludes with the observation that 'a WTO compatible set of environmental sustainability criteria would not be easy to craft. At the very least the package would have to be non-discriminatory, scientifically based and only implemented after serious negotiations with potential suppliers'. We offer a way into the WTO-compatibility debate by looking at the advice the TBTA offers on how to design standards. This is informed by our interview data, which includes an interview with one of the chief architects of the EU sustainability criteria.

To recap some of the foregoing first, Article III of the GATT applies to biofuels. As a result, if there is to be a general exception either sought or identified for biofuels, it is via Article XX, in particular XXg – subject to conformity with the principles of national treatment and non-discrimination. This point remains moot, however, until or unless an action is brought against a country's biofuels policy to test this in law. This is despite the fact that the preamble to the Agreement Establishing the WTO refers to boosting trade 'while allowing for the optimal use of the world's resources in accordance with the objective of sustainable development, seeking both to protect and preserve the environment and to enhance the means for doing so'.

We turn now to the EU sustainability criteria. These are an example of product standards and thus fall under the remit of the TBTA. As noted, whilst the GATT sets out what can and cannot be done, the TBTA offers guidance on how to do it. Specifically, it provides guidance about how to avoid measures which could affect trade by creating 'arbitrary or unjustifiable discrimination between countries' in their implementation and impact. The TBTA also makes the distinction, crucial in the case of the sustainability criteria, between mandatory technical regulations and voluntary standards.

Advice offered by the TBTA includes the recommendation that, in drawing up standards countries should, where possible, draw on existing international standards and agreements. If that is not possible or appropriate, the country should involve other countries openly and actively in discussions, prior to the drawing up and implementation of new standards. Following such advice makes practical sense, beyond simply doing what the TBTA says, because it makes it less likely another WTO Member will bring an action subsequently against those sustainability criteria. We can identify three distinct phases to the development of the sustainability criteria in the EU, each of which, in some way, reflects the two options identified above. The first involved open engagement with other parties in the development of new policy instruments. The second related to the content of the sustainability criteria. The third relates to the ongoing process of implementation, which involves ensuring the sustainability criteria are respected. For this, an existing set of institutions and processes have been utilised, albeit with adjustments where necessary.

First came the process of negotiating the criteria. To quote Ackrill and Kay (2011b: 558) at some length:

In accordance with Article 2.9.4 of the TBT, the EU was required to give other WTO Members a reasonable period of time in which to comment on its proposed [environmental sustainability criteria] 'and take these written comments and the results of these discussions into account'. This it did on 11 July 2008,[16] when it published the proposals for what would become the RED. It then sent an Addendum to inform WTO members that the legislation had been adopted; not in April 2009 when the RED was agreed, nor June when it was published; but in June 2010[17] when the Commission published its Communication 'on the practical implementation of the EU biofuels and bioliquids sustainability scheme and on counting rules for biofuels'[18] .... There is no requirement the EU take any comments into consideration, but [evidence indicates] the EU did discuss widely (and interviews with non-EU stakeholders confirm their active engagement in this process).

In short, the EU adopted a process of negotiating the criteria which was open and which engaged with other countries. It is also interesting that, as one interviewee told us, prior to discussion of the sustainability criteria, other countries had not really been engaged. With this aspect of policy, however, the significance to them was recognised and so particular effort was made to engage with them over the sustainability criteria.

The second dimension is what actually went into the criteria. As important – reflecting also on what other WTO members' views might be – is what was left out of the criteria. Interviews confirmed that the design of the criteria was undertaken with WTO factors explicitly in mind. As a result, labour standards are mentioned in the criteria, but are not included as a pre-condition for sustainability. The European Parliament, for example, supported compulsory labour standards. Moreover, some experts (notably Charnovitz *et al.*, 2008) have argued labour standards can be designed and implemented in ways consistent with WTO rules. The view taken by the Commission, however, was that such rules would overstep some peoples' 'red lines'. This risked an action in the WTO which, it was felt, could threaten the entire biofuels policy, not just the detail of the sustainability criteria.

Instead, a process of two-yearly reporting on the implementation of the policy by the European Commission was put in place. This would consider, for example, whether major source countries for biofuels in the EU were implementing a range of International Labour Organisation Conventions, the Cartagena Protocol on Biosafety and the Convention on International Trade in Endangered Species of Wild Fauna and Flora. Further, the reports will highlight examples of good practice. This soft approach to seeking to improve policy implementation is similar to what the Commission does already in a range of domestic policy areas, utilising the Open Method of Coordination.[19]

Related to this is the fact that fuel companies are given an incentive to blend biofuels in transport fuel by means of fiscal incentives, rather than have a legal obligation placed on them. A legal obligation would turn implementing the sustainability criteria into a regulation rather than a standard, as defined in the TBTA. This would, in turn, imply that fuel companies would only be able to import biofuels derived from sustainably produced feedstocks, which would represent a trade barrier at the border. By the setting of a voluntary standard, in TBTA-speak, fuel companies do not have to import only (or purchase domestically) and blend sustainable biofuels; it is just that if they choose to purchase non-sustainable biofuels, they will not receive the fiscal incentives.

In the EU context, another very important part of the policy is that, in implementing the sustainability criteria, member states are not allowed to add further conditions to the measures agreed collectively. Given those EU criteria were designed to be WTO-compliant, this restriction ensures member states cannot undermine this unilaterally. Moreover, member states must produce National Action Plans to show how they will deliver on the sustainability criteria (Switzer and McMahon, 2010: 6). These provide a point in the process where this can be monitored. This restriction on unilateral action by member states also helps address another emerging problem. There is a global proliferation of sustainability standards which could contradict each other, cause confusion and undermine the effectiveness of the policy (Desplechin, 2010).

The third dimension is the implementation of the criteria. Here, as introduced in the previous two chapters and analysed in greater detail below, implementation of EU policy requires production of the biofuels feedstocks to be certified, in order to seek to ensure the resulting biofuels conform with the sustainability criteria. The TBTA recognises a role for NGOs to be involved in standard-setting. The EU approach has been to allow a wide range of bodies to propose certification procedures, which are then approved if they are deemed suitable for determining the compliance of feedstock production with the sustainability criteria. As of late 2014, 19 separate certification schemes have been approved.[20]

Some of these certification bodies have come new to such certification, for example, Ensus has established a scheme that applies solely to the supply chain delivering ethanol derived from feed wheat at its factory in the UK. Other approved schemes are operated by organisations who have long experience in certification processes. That is, the EU is tapping into an existing network of organisations and established sets of practices and standards – as suggested in the TBTA. Examples include Bonsucro[21] (whose work relates to sugarcane production), the Roundtable on Sustainable Palm Oil (RSPO)[22] and the Roundtable on Responsible Soy (RTRS),[23] and the Roundtable on Sustainable Biomaterials (RSB).[24]

Taking Bonsucro as an example, it already operated a scheme which contained five core principles,[25] each of which is divided into multiple indicators. For the EU, a sixth category was added specifically to ensure compliance with the RED and FQD. This contains two indicators, to seek to ensure full compliance with the EU ESC: to monitor global warming emissions with a view to minimising climate change impacts; and to protect land with high biodiversity value, land with high carbon stock and peatlands. A seventh section to 'Bonsucro EU', addressing 'chain of custody requirements', also makes explicit reference to the demands of the EU RED.

It is also important in relation to the TBTA that these NGOs operate in an open and transparent manner, and they have broad membership, including non-governmental representation, from all relevant producing countries. This process of 'roundtabling', Ponte (2013) argues, is actually extremely important in producing a system that is transparent, accountable and inclusive. That said, he argues it is also complex and can be time-consuming – as a result of which, opportunities have arisen for other actors to enter this policy arena that utilise less positive processes and deliver less desirable outcomes: 'Commercially-oriented initiatives are generally less democratic, leaner, quicker, and more attuned to industry interests' (Ponte, 2013: 9).[26]

As the foregoing general discussion has revealed, however, the links between biofuels policies and WTO trade rules remain untested. Interviews in Brazil, for example, confirmed the Brazilian government engaged actively with the EU as the latter prepared its sustainability criteria, but they also revealed that it is taking a hands-off approach regarding the decisions of Brazilian private sector companies to join these schemes. The reasoning is that it wishes to remain neutral whilst observing their implementation. If WTO rules are violated in implementation, the government wishes to ensure it has not given the impression of approving of such schemes by playing any part in firms' individual decisions to participate.

The transboundary implementation of domestic policies

The foregoing analysis leads us very neatly into the second major issue with biofuels policies: with the sustainability criteria, policy-makers have designed an element of biofuels policy which, in implementation, operates across national boundaries. Because this is linked with sustainability criteria, it is seen most clearly with EU policy. We therefore use the EU biofuels case to illustrate this policy challenge. It should be made clear from the outset, however, that this is not only a feature of biofuels policies: with increasing globalisation and the weakening of the external sovereignty of the nation-state since the end of World War II, this is an ever-growing policy challenge in the 21st century. What we discuss here therefore links back to the analysis introduced at the start of Chapter 6.

What, then, is sovereignty, and how is it being affected by globalisation? One challenge is that sovereignty tends to get discussed rather than defined. Kobrin (2009: 185–186) distinguishes between internal and external sovereignty thus. 'Internal sovereignty defines the legitimization of the state vis-à-vis competing domestic claimants . . . [whereas] . . . [e]xternal sovereignty is a more amorphous concept'. Internal sovereignty thus confirms the primacy of authority a state has over its citizens. As for external sovereignty, its presence would be marked by there being no authority higher than a given state: actions over a nation-state – by another state, or indeed by a non-state body – would not be feasible. States would be equal to each other, rather than some being subordinate to others – subject only to the condition that each state is recognised by others.

Related to the concept of external sovereignty is that of autonomy, defined (for example) by Kobrin (2009: 186) as 'a political idea which implies that a state can make its own decisions about how it will deal with internal and external problems'. This grounds sovereignty in practical questions of business policy and regulation. This is where globalisation enters the discussion, insofar as economic and business decision-making is concerned. For example, Kobrin argues that the growth of multinational companies (MNCs) has eroded a state's policy autonomy. Matthews (2012: 282) refers to 'the paradox of state capacity', wherein states have sought to develop new forms of policy capacity, while at the same time transferring key control levers to a range of semi-independent organizational forms'. Kobrin (2009: 198–199) goes so far as to argue that 'globalisation is weakening territorial sovereignty to the point where economic and political governance based primarily on geographical jurisdiction may no longer be viable'.

To recall terms introduced at the beginning of this chapter, this represents a shift from a hierarchical relationship, of states over non-state actors, to a flatter 'network', where states must work with non-state actors in a more consensual style (Wolf, 2008; Richardson, 2012). This is crucial to biofuels because it stands as a case where policy-makers design a public policy, but one requiring private actors for its successful implementation. It involves, as indicated earlier, MNCs, NGOs and international organisations, notably the WTO, operating in a space over which the EU enjoys no jurisdiction or direct control. Ultimately, this rejects the traditional view that the same public authority (the Westphalian nation-state) exerts control over policy design and policy implementation.

Richardson (2012: 311–312), quoting the work of Héritier and Rhodes, argues that 'in a highly complex society, with problems extending across borders, central actors are unable to muster the knowledge required to shape effective instruments of interventions. They depend on the expertise and knowledge of private and local actors. In these conditions, centralized and hierarchical steering is doomed to failure'. This demonstrates that we are talking about situations which go beyond trade – it relates to hands-on activities, 'interventions', in other states: this is our transboundary policy implementation problem. The clearest example of this arising in biofuels policy is with EU sustainability criteria.

The earlier analysis referred to the policy design stage of these criteria and, as such, we were more interested in process than content. We now need to look at these criteria in detail, to understand the nature and extent of the implementation challenge, as it extends beyond national borders. These are set out in Article 17 of the RED, and Article 7b of the FQD. Any biofuel can legally be marketed in the EU, but only those which comply with the sustainability criteria can count against the biofuels usage mandate and be eligible for fiscal concessions to the (mainly multinational) companies buying and blending the biofuels into the fossil fuel.

Eligible biofuels must deliver GHG emissions reductions over fossil fuels – of at least 35 per cent initially (from 2013 if the production facility was operating before 2008), at least 50 per cent from 2017. From 2018, biofuels produced in plants which began production in 2017 must deliver savings of at least 60 per cent. Details are provided in the legislation regarding how to calculate GHG emissions reductions. An additional criterion which appears in Article 7a of the FQD, but not the RED, is that fuel suppliers must deliver a 6 per cent reduction in life-cycle GHG emissions from the fuel they supply by 2020, relative to average fossil-fuel life-cycle GHG emissions in 2010. Whilst this reduction can come about by other means, within the fossil-fuel supply chain, it is expected that most will be achieved through the incorporation of biofuels into fossil fuels (see Swinbank and Daugbjerg, 2013).

In addition, as shown in Box 3.1, a series of exclusionary criteria are set out, identifying certain types of land with a specific function or status before 2008: land which cannot be used, so as to preserve biodiversity, and land which acts as a significant carbon store. This requires the production of agricultural/biofuels feedstocks to be monitored, to ensure it conforms to these criteria – feedstocks that could potentially be produced anywhere in the world (although most of the best-performing feedstocks tend to be tropical products). Recalling Richardson's argument, above, policies 'depend on the expertise and knowledge of private and local actors', as a result of which 'centralized and hierarchical steering is doomed to failure'. Globalisation and the shift from hierarchy to network has, for biofuels, resulted in a range of actors now deeply embedded in the policy implementation process. We have discussed in detail in this chapter the role of the WTO. We see above that (mainly private sector) MNCs – the fuel companies – must deliver transport fuel which reduces emissions and, if they wish to receive the fiscal incentives in place, must deliver fuel which contains biofuels produced in accordance with the sustainability criteria.

As the earlier discussion and related link to the European Commission's website indicate, only a minority of schemes approved to certify feedstock production as 'sustainable' are operated by fuel companies. Most schemes are undertaken by private sector NGOs, some of whom were shown earlier to have considerable experience in certification more generally and who have been brought into the biofuels policy network. These embody the 'private and local actors' to which Richardson refers. They also embody, along with the MNCs and the shadow cast by the WTO, the complexity policy-makers now face when implementing policies which require management ('governance') beyond national borders. It is important to note that such a network does not involve policy-makers from the different countries working together, but the policy-makers in one jurisdiction working with private actors who are operating in other countries. Thus notions of autonomy and sovereignty are changing with globalisation – developments which are well illustrated by the nature of biofuels policies.

Other trade-related concerns for biofuels

In this final section, we consider just some of the other trade-related issues regarding the potential development of a global biofuels industry, to quote from the title of Loppacher and Kerr (2005). One point these authors make concerns subsidies – in particular subsidies in the context of the WTO Agreements as discussed earlier. They – and many others – point out the need for subsidies to be provided for various renewable energies, in order to allow them to compete with fossil fuels. One issue is the very low cost of extracting crude oil in some countries, whilst decades of investment have helped build up scale and efficiency.

Loppacher and Kerr, along with several other papers cited in this chapter, raise concerns about the WTO compatibility of renewable energy/biofuel subsidies and whether or not an action could be triggered in the WTO. There is another aspect to this subsidies debate, however – because the foregoing, by looking only at subsidies for renewables, only looks at part of a bigger picture. And that bigger picture shows us hundreds of billions of euros of subsidies being given to fossil fuel industries each year.

A recent news article[27] quotes from a 2009 meeting of the G20[28] group of countries: 'the inefficient subsidization of fossil fuels supports wasteful behaviour, complicates investments in clean energy sources and undermines efforts to fight the dangers of global warming'. They indicate that, just a few years on, fossil fuel subsidies are, depending on the exact calculation method, somewhere between €400 billion and €2.6 trillion. This compares to estimates from the International Energy Agency for renewable energy subsidies, that in 2011 the figure was $88 billion, roughly €66 billion. This raises two issues. First, the consistency of treatment of fossil-fuel and renewable energy subsidies in the WTO, were the latter to be subject to an action, and, second, the challenges facing renewable energies in trying to compete against fossil fuels in receipt of such levels of subsidy.

Related to this, Loppacher and Kerr raise the issue of trade policies which seek to block products which include biotechnology-related inputs somewhere in the supply chain. In particular, they argue (on page 15) that biotechnology is a 'critical tool in reducing the cost of producing biofuels so that they become more competitive with petroleum-based fuels'. In the EU in particular, however, there has been long-standing and deep-seated opposition to genetically modified organisms (GMOs) and GM crops. As a result, conflict between the EU and, in particular, US led to a case in the WTO, brought in May 2003.[29] In it, it was alleged that the EU had imposed a *general* moratorium on the approval of biotech products, introduced measures affecting the approval of *specific* biotech products, and safeguard measures prohibited the import and marketing of specific biotech products. Even though the findings broadly supported the US, the situation remains uncertain, with profound differences remaining (for an outstanding analysis of the case and the related issues, see Pollack and Shaffer, 2009).

There is a twist to this debate, however. Whilst the WTO, clearly, sees concerns over GMOs manifested as policies such as the EUs, as a distortion of trade, other international agreements take a different view. In particular, The Cartagena Protocol on Biosafety to the Convention on Biological Diversity[30] allows for the use of the 'precautionary principle' in the face of unknown risk. In so doing, it is also re-stating a principle set out as Principle 15 of the 1992 Rio Earth Summit (discussed in Chapter 3).[31] The EU defines the precautionary principle thus on its Europa website:

> The precautionary principle enables rapid response in the face of a possible danger to human, animal or plant health, or to protect the environment. In particular, where scientific data do not permit a complete evaluation of the risk, recourse to this principle may, for example, be used to stop distribution or order withdrawal from the market of products likely to be hazardous.

It is also ambiguous whether this definition covers GMOs – it all depends on one's view as to whether the scientific evidence does, or does not, provide the means for a complete evaluation of the risk attached to a given 'specific' GMO, regardless of the general stance taken on GMOs generally. We leave this point here as this debate is one that will run and run.

Another issue that is causing difficulties with biofuels trade and its relationship to international trade rules is that there is no simple, clear, codification of ethanol and biodiesel in the Harmonised Commodity Description and Coding System (HS). Ethanol is classified under HS Chapter 22 ('beverages, spirits and vinegar'), a classification which is complicated by sub-divisions into either undenatured ethyl alcohol or various types of denatured ethyl alcohol. Biodiesel, meanwhile, falls under HS Chapter 38 ('miscellaneous chemical products'). As noted earlier, agricultural products are treated differently in the WTO in some regards – and this classification means that ethanol falls under the AoA, although, as Howse *et al.* (2006: 11–12) point out, if the WTO Members wished to, they could remove ethanol from the AoA. There is a further problem with this, which makes it very hard to analyse trade in biofuels: these HS categories refer to ethanol and biodiesel *regardless* of their end use, as a fuel or for other purposes.[32]

Whilst trade in biofuels is difficult to analyse for this reason, there is one oddity in biofuels trade which has emerged in recent years and which merits comment. In economics, the phrase 'intra-industry trade' (IIT) describes an increasingly common trade pattern, whereby two countries simultaneously export and import the same good with each other. There are various reasons for this. For example, a country might import seasonal fruit and vegetables for part of the year, and export them the rest of the year – and if trade data are for the whole year, it looks like they are being traded at the same time. Much more common is the case where products are very similar, but not identical. Many countries import and export cars – because the product 'cars' embraces lots of variety. Therefore IIT provides consumers with lots of variety, lots of choice.

And so it is that, in recent years, we have seen this pattern emerging in ethanol trade between Brazil and the US – except that in this case, the IIT trade pattern is driven by policy differences. Earlier in the chapter we discussed the notion of 'like products' as defined by the WTO. The basis for this is related primarily to the uses to which the product can be put. One of the challenges for countries seeking to ensure policies are WTO-compliant is that, as we saw earlier, there are no policies which prevent trade in biofuels produced unsustainably. Hence, in the EU and US, the use of fiscal incentives to try to ensure fuel companies blend only biofuels produced sustainably in the petrol and diesel they sell. Thus it is that different policies have created IIT in ethanol.

In this case, however, there are two countries but three sets of policies – Brazil's policy of blending ethanol (normally) to 25 per cent in petrol as well as offering E100; US policy which has promoted a large expansion of corn ethanol, delivering at least 20 per cent GHG emissions savings (but typically nowhere near the 50 per cent plus that Brazilian sugarcane ethanol can deliver); and the policy of the California Air Resources Board (CARB), the Low Carbon Fuel Standard Program (LCFS).[33] This aims to reduce the carbon intensity of transport fuels used in California by 10 per cent by 2020. The policy was agreed in 2009 and came into effect in 2011. Controversially at the time, because it included estimated land-use change effects (via flat-rate values, as the EU is currently also proposing), it effectively excluded Mid-Western US corn-based ethanol (on the extant measure of carbon intensity) – even though the ILUC factor for corn ethanol, 30 g/MJ, was below that for sugarcane ethanol (46 g/MJ); and despite which, Brazilian sugarcane ethanol still met the CARB standard.[34]

The upshot of this was that from 2011, Brazilian sugarcane ethanol was being exported to California, to satisfy the demands of the LCFS. Meanwhile, fluctuations in Brazil's sugarcane harvests, and sugar and oil price fluctuations, resulted in shortages of ethanol production and imports being bought from the US, to ensure adequate supplies to blend into petrol at E25 (periodically in recent years reduced to E20), even if rising ethanol prices were suppressing demand for E100. One particular article about this 'Ethanol Shuffle'[35] points out that, in 2011, corn ethanol was cheaper to produce than Brazilian sugarcane ethanol (ranging between $1 and $1.50 per gallon in late 2010 and early 211). This article also estimates that the transportation-related emissions are double, compared to the situation where US corn ethanol supplies US demand and Brazilian ethanol supplies Brazilian demand.[36]

Thus Californian policy focuses on the GHG emissions arising from the production of the biofuel, by which measure Brazil's sugarcane ethanol is the superior product, and which renders US corn ethanol ineligible. Brazilian policy, meanwhile, makes no reference to the GHG emissions performance of the ethanol it blends into petrol, as a result of which it is perfectly free (policy-wise) to import US corn ethanol, despite its lower GHG emissions-reduction performance. Further, the blendwall is creating demand-side constraints in the US market, in particular with its main domestic product – corn ethanol. The result is the ethanol shuffle.[37]

Our final issue to consider in this chapter is another trade oddity that occurred between the US and Brazil, relating to aspects of US policy analysed in Chapter 4. As explained there, the US for a long time implemented an ethanol tax credit – formally, the Volumetric Ethanol Excise Tax Credit (VEETC). Because the VEETC was designed as a domestic policy measure, but domestic blenders were unable to distinguish between the US and imported ethanol, the government imposed a tariff on imported ethanol to offset any advantage the VEETC may confer on non-US suppliers of ethanol. The headline figure was a tariff of 54 cents per gallon, but an additional 2.5 per cent ad valorem component took the effective tariff up to 57 cents per gallon (Devadoss and Kuffel, 2010: 477). As analysed in Chapter 4, the VEETC and the tariff were allowed to expire at the end of 2011.

For several years prior to 2011, Brazil threatened to bring an action against the tariff in the WTO, and maintained diplomatic pressure on the US over this matter. Interviews in Brazil confirmed that Brazilian officials made representations about this to US officials at every opportunity. Given Brazil's success at the WTO in the Upland Cotton dispute, brought against US cotton policy,[38] as well as success (jointly with Australia and Thailand) against EU sugar policy,[39] it was developing a credible track record of successful cases against agricultural and related policies in the WTO. By 2011, senior US officials recognised the WTO would not uphold the policy,[40] which begs the question – why did Brazil not bring the action via the DSP?

Whilst our interviews did not produce a definitive answer to this question, they did raise one interesting possible factor. Notable in this was what Brazil was able to do instead, in order to gain access to the US market. Under the Caribbean Basin Initiative (CBI),[41] ethanol from Caribbean countries could be exported tariff-free into the US. Under the terms of the CBI, however, products that had been processed in one of the CBI countries could enter the US tariff-free. As a result, a trade developed whereby Brazil would export ethanol to a CBI country, where it would be processed (a little bit) then shipped to the US. Typically Brazil would export hydrous ethanol, which would be dehydrated in the Caribbean, then exported to the US where, as anhydrous ethanol, it could be blended into petrol.[42]

It was therefore suggested to us that one reason for Brazil not taking the US to the WTO over the tariff was that, whilst the US was not happy about the CBI-routed trade, an acceptance emerged that resulted in Brazil not pursuing the WTO case, whilst the US did not try to tackle the CBI issue. As explained in Chapter 4, however, a range of factors, including the domestic fiscal cliff in the US, led to the non-renewal of the VEETC and, with it, the abandonment of the tariff. Whilst it was now once again more profitable for Brazilian ethanol to be exported directly to the US, significant business was lost from some CBI countries. This, along with factors such as rising prices of Brazilian hydrous ethanol, has resulted in the scaling back of the Caribbean ethanol processing industry.[43]

Ultimately, as explored in this and the preceding two chapters, biofuels policies were conceived as domestic policies and have been implemented as such. There are international fora (such as GBEP) where national policy-makers share information, and there are international organisations (such as the WTO) which provide different platforms by which national policies and policy-makers can interact. There are inevitably international dimensions to policies and commodities that are traded, as we have seen in this chapter. Yet, as demonstrated in Chapter 7, the nature of domestic policy-making has, thus far, attenuated any impact of the international on the domestic. And despite, for example, the avowed aim of the Brazilian government to create a global market in biofuels, there is most definitely not yet a global biofuels policy regime (Bastos Lima and Gupta, 2013).

Conclusions

Biofuels policies were set up initially as domestic policies. As Chapter 7 set out, several elements in the initial policy designs have helped keep the politics of biofuels policy primarily at the domestic level. However, as this chapter has shown, the external dimensions of biofuels policies extend far and wide and that in open economy politics the external dimension is always present to some degree. Even domestic policies, insofar as they can affect trade, must seek compliance with the international rules on trade that all WTO members have signed up to. Designing new policies from scratch, as has happened with many elements of biofuels policies, is, however, a double-edged sword. On the one hand, starting from scratch means being able to design policies which, at least on paper, look like they should conform with WTO rules. On the other hand there are many policies whose WTO-conformity would be determined only when, or if, a case was to be brought against it by another WTO member. This can create policy and political uncertainties.

Following on from the themes of Chapters 6 and 7, in this chapter we have seen how these international dimensions to biofuels policies have raised questions at both the design and implementation stages. Whilst many biofuels policy instruments (subsidies, tariffs and so on) are familiar in the WTO context, other policies and policy instruments, notably sustainability criteria for the production of (land-based) biofuels feedstocks, are either new, or are used in new contexts. Thus whilst several certification bodies have certification schemes which apply to different products as well as biofuels, there is a fundamental difference in WTO law between giving a consumer a straight choice 'on the shelf' between certified and non-certified products (for example, Fairtrade and non-Fairtrade products), and biofuels, where governments want only certified biofuels to be used, but who are limited in their possible actions because of those WTO rules as they apply to non-certified products.

The rules and mechanisms put in place to implement and ensure respect for sustainability criteria, in turn, raise another significant issue that is particularly pertinent in the 21st century – how governments attempt to implement and manage policies created domestically, but which have a reach far beyond the borders within which their sovereignty is confined. Biofuels policy, especially in the shape of EU sustainability criteria, provides an excellent example of how governments are trying to 'manage' globalisation (Jacoby and Meunier, 2010).[44] EU policy involves coordinating and collaborating with a range of types of private sector actor to work towards what, it is hoped, are goals and outcomes of common interest. This may involve the use of, for example, fiscal incentives to MNCs such as fuel companies, but all such mechanisms still leave space for choice. One reason for this is that shadow cast by the WTO. This is made more complex because legally binding regulations and voluntary standards have different legal implications under WTO rules and principles.

9
Biofuels Policy Challenges

Introduction

Biofuels policies – and biofuels policy-makers – face considerable challenges in delivering on existing policy goals, let alone looking to expand biofuels markets. In this chapter, we focus on three of these challenges, two of which have been central to the biofuels debate in recent years, the third notable for being largely absent. The first two policy challenges – food v. fuel and ILUC – arise from the implementation of biofuels policies. Moreover, they have similar underpinnings, in that they result from the complexity and interconnectedness of the markets wherein biofuels are produced.

The third factor is the possible role played by end-consumers in both the design and implementation stages of biofuels policies. Thus far, policies promoting biofuels production and use, most notably in the EU and US, have developed without any particular role for the end-users – hence their absence from the analyses presented in Part I. Yet, as implied there, biofuels have been represented frequently in the media, and consumers and potential consumers tend to have views on biofuels on that basis. We therefore explore the links between biofuels and biofuels policies, consumers and media representation.

The question of food v. fuel arises, as introduced in Chapter 1, because first generation biofuels are derived from agricultural commodities which can be used either as food for humans, or feed for animals (which, in turn, either produce, or themselves become food for humans) and which are grown on land that could instead be used for producing food or feed.[1] This affects the supply of agricultural commodities for food use. Second generation biofuel feedstocks, by definition, avoid the first of these two problems, but do not avoid the second entirely,

180

as the production of some feedstocks also requires land. ILUC arises because changes in the demand for and supply of feedstocks for biofuels production affect commodity prices, which in turn results in farmers – potentially anywhere around the world – altering their production and land-use decisions to produce and supply affected food-related markets.

Both food v. fuel and ILUC arise, therefore, because of the complexity of overlapping markets. If a biofuel is derived from a feedstock that has been diverted from use in the production of a food product, the supply of that product into the food product market will change. This might affect the price of that food product. This, in turn, might impact consumers, if the rise in price of the food product is passed all the way through to the retail end of the supply chain. Meanwhile, that initial shift in the use of the feedstock, and its possible impact on agricultural commodity market prices, will also affect producers, and it might, in turn, also result in producers elsewhere changing their production decisions as a result of price changes, thus potentially triggering DLUC and ILUC.

But we only say 'might'. The biofuels-related production and land-use decisions described in this book are just some of the myriad economic exchanges and decisions occurring around the world, continuously, which can affect commodity prices, food prices and farmers' production decisions. To isolate the effect of A (using feedstocks for biofuels) on B (the price of that feedstock in different commodity markets), on C (consequent on B, the price of that feedstock in different food markets) and of A *and* B on D (the production decisions of thousands, possibly millions, of farmers around the world, regarding that feedstock and other crops that could be used for food and/or biofuels) is incredibly difficult.

In order to attempt to isolate the impact of biofuels in both the food v. fuel and ILUC debates, economic models are used. The task of modelling such complexity is formidable. In the case of ILUC, this is compounded by economists trying to represent something that is not actually visible, as explained later, as a result of which we do not have an empirical baseline against which the judge the estimations obtained from economic models. Such complexities of market and policy interactions mean that, in trying to estimate the magnitude of a given 'downside' to biofuels, we cannot even be certain that biofuels have been the cause of that downside – examples of which will be given below. Further, if biofuels have contributed to the downside, we also need to try to estimate how much of that downside has been caused by biofuels, rather than the many other possible contributory factors.

Alongside the similarities between the food v. fuel and ILUC debates, there is also an inherent contradiction. ILUC sees producers around the world bring land into agricultural production as a result of the price-related consequences of agricultural feedstocks, produced elsewhere, being shifted out of food production and into biofuels production, an act which releases GHGs. The very phrase 'food v. fuel', however, implies that we are dealing with a zero sum game: that the total available land for the production of food crops is fixed, more or less, and therefore represents a shift in land *use* from producing feedstocks for food, to producing feedstocks for biofuel, of more or less exactly equal and opposite magnitude. If we are able to bring land into agricultural production, however, to trigger ILUC effects, any food v. fuel conflicts would, as a minimum, be lessened, and perhaps even negated. We explore this issue further below.[2]

Amongst the issues arising from using agricultural feedstocks to produce biofuels, one of the most widely used quotes comes from Jean Ziegler, the UN Special Rapporteur on the right to food. In 2007, he declared 'it is a crime against humanity to convert agricultural productive soil into soil which produces food stuff that will be burned into biofuel'.[3] In the same speech, however, he also stated that already we produce 'enough food to feed every child, woman and man and could feed 12 billion people, double the current world population, according to the UN Food and Agriculture Organization (FAO)'.[4] Since 2000, roughly 2 per cent of global agricultural land has been given over to the production of feedstocks for biofuels. It is therefore not at all clear how, if we are already producing enough food for 6 billion, and could produce enough for 12 billion, the shift of relatively so little land into producing feedstocks for biofuels be the cause of the food security challenges facing the world. If, moreover, by Ziegler's own reckoning we are already producing enough food in total for the total global population, the problems related to food security must lie elsewhere, rather than simply in terms of questions about supply, which is the focus of food v. fuel (see below).

Meanwhile, it has been estimated that between a third and a half of all food production worldwide is wasted.[5] In rich countries, this tends to be located at the consumer end of the food chain. Consumer preferences over the appearance of fruit and vegetables, supermarket sell-by dates, the failure to consume perishable foodstuffs in time, unconsumed food at restaurants and so on, all add to this figure, often with food simply being thrown away. In poor countries, the problem tends to be located closer to the production end of the supply chain. Inefficiencies with agronomic practices, harvesting, storage, transportation and processing can all lead to significant loss of food between field and fork.

Since people are starving despite the fact that, as Ziegler acknowledges, there is enough food being produced, what exactly is the problem? A key part of the answer lies in the distribution not only of production, but also of the ability to access food. A detailed analysis of this is well beyond the scope of the present book, but the idea is well-captured by Amartya Sen's concept of entitlements (Sen, 1981; Drèze and Sen, 1989). The essence of entitlements is that individuals can gain title over food in two ways – through what they grow and what they buy. Unequal entitlements are reflected not only in the uneven distribution of food self-production, but in the unequal distribution of 'purchasing power', the ability to acquire food by those who, for whatever reason, do not produce food for themselves. This is not to suggest that, in localised situations, biofuels cannot impact on food security; just that the role attributed to biofuels in broader debates over global food security appears to be over-stated. We shall also see this below, when we look in more detail at one possible channel by which biofuels can affect food security: via higher food prices. Moreover, we also look at the wider implications for this debate arising from the work of one academic, who has asked the question 'what is the right price of food?' (Swinnen, 2011).

In the remainder of this chapter, we explore some of the complex issues behind the food v. fuel debate and ILUC, before concluding with a look at the demand-side of biofuels markets. Our aim here, as it has been throughout this book, is not to try to provide the reader with definitive answers. Indeed, we would argue that, when looking at the complexities inherent in biofuels markets and policies, simple clear answers do not exist. Instead, we attempt to set out the complexities, explore some of the research undertaken, and thus aim to help equip the reader with a toolkit of ideas that can be used to understand, to interpret and to confront the debates around biofuels policies – in particular in relation to what defines biofuels done well or done badly.

In this, we must remember a very simple but absolutely critical point that is often overlooked – the focus of these debates is not actually the biofuels themselves, but the feedstocks from which the biofuels are derived; hence the wide acceptance of the (expected or hoped-for) superiority of advanced biofuels over the first generation biofuels which dominate the market currently. We thus, in this chapter, focus on both ends of the biofuels supply chain. Food v. fuel and ILUC debates focus around the upstream end and the production of the inputs (see also DeCicco, 2013), whilst discussion of consumers focuses more on the downstream end – although several of the concerns consumers have relate directly to those upstream issues.

Food v. fuel – A real or false dichotomy?

The idea behind the food v. fuel issue is very simple – by diverting feedstocks from food production to biofuels production, the availability of agricultural commodities for food falls, the supply of food falls and prices rise. The combination of falling supplies and rising prices harms consumers – and, given the global interconnectedness of markets, this affects consumers in countries around the world. Moreover, this will affect poor consumers harder (in both rich and poor countries), because the poor spend a higher proportion of their incomes on food than the rich.

In order to explore the issues underpinning the food v. fuel debate in more detail, we shall consider just a small part of the considerable literature available on this topic.[6] First, we look at some of the work conducted which explores, in particular, the food-price impacts of biofuels production. We then look at work, the focus of which has been to understand the causes of rising prices and rising price volatility in recent years, in particular the price spikes of 2007–2008 and 2011. In so doing, we shall get a clearer picture of the wide range of factors which affect world food prices. Much of this work is highly technical in nature – with some research papers focusing as much on the econometric methods employed and the consequences for estimation, as the results themselves. In what follows, we try to offer a representative cross-section of work, providing points of comparison, similarity, difference and development. We begin with a few papers which focus on single countries, before opening out to look at multi-country and/or multi-commodity studies.

Starting with Brazil, Serra (2011) finds strong volatility links between energy and food markets, but that whilst crude oil and sugar prices influence ethanol prices, Serra found no evidence that ethanol prices influenced food prices. She attributes this to the expansion in land area used for sugarcane cultivation. In contrast, Serra *et al.* (2011: 271), using a shorter data period (July 2000 to February 2008, rather than November 2009 in Serra, 2011), find that '[s]hocks to the oil *and ethanol* markets also cause instability in the sugar markets.' (emphasis added). That said, shocks transmitted from ethanol to sugar markets tend to be abrupt and relatively short-lived. This may reflect the relatively free market in Brazil and the existence of refineries capable of producing both sugar and ethanol, whereby price shocks can feed through to supply and demand of the two products relatively swiftly.

Busse *et al.* (2010) look at a similar issue in the EU, in their case biodiesel and vegetable oil in Germany. They find evidence of a strong impact of the price of crude oil on the price of biodiesel, and of the price of biodiesel on the price of rapeseed oil and soya oil. That said, the nature and strength of these relationships varied over time, for a variety of reasons, including changing German policy support, the food price spikes of 2007–2008 and imports of US biodiesel. Interestingly, they find that 'increasing imports of biodiesel have driven the EU biodiesel market from an insulated market to an internationally contested market' (Busse *et al.*, 2010: 23). This is likely to increase price transmission effects. That said, reflecting the multiplicity of factors affecting commodity prices, they argue that the rise in vegetable oil prices in Germany during 2007–2008 was caused more by shifts in international vegetable oil prices than by biodiesel prices.

Kretschmer *et al.* (2012) is a useful reference, as it provides a survey of the work looking at the impact of EU biofuels policy on agricultural commodity prices (see Table 9.1). Their main findings can be summarised as follows. The estimated impacts of EU biofuels policy on the prices of different commodities vary enormously between studies – even for the same commodity. That said, the impact on oil crops is greater than for cereals and sugar – a result consistent with the relative importance of biodiesel and ethanol in total EU biofuel (and fossil-fuel) consumption. For comparison, the results of studies with a global focus find that the impact on ethanol and biodiesel is either largely balanced, or greater on ethanol, by far the more extensively consumed biofuel globally.

That said, whilst the estimated impacts are positive, they are 'not massive' compared with recent spikes in world commodity prices, although 'they are not negligible either'. (Kretschmer *et al.*, 2012: 49). The authors' final comment is also important: EU policies do not operate in isolation. Rather, policies implemented globally can work together to affect commodity prices. Ultimately, there is little evidence from Table 9.1 to suggest policy interaction is mutually reinforcing, increasing the impact on commodity prices.

Turning to some of the studies which look at the US, Qiu *et al.* (2012) find that the price volatility of oil, ethanol and corn comes mainly from own-product demand shocks (corn is affected by shocks to corn and so on) – there is no evidence that shocks to the oil, petrol and ethanol markets spill over to affect the corn market. Moreover, the impact of a

Table 9.1   Price effects of EU biofuels policy, and of global biofuel mandates

Feedstock	Range of price effects (per cent)	Comments
Studies that focus on the effects of EU biofuel policy		
Oilseeds	8–20	
Vegetable oils	1–36	
Oilseeds	9–20	
Cereals/maize	1–22	One study estimates 22 per cent for maize. All other studies estimate increases of no more than 8 per cent, for wheat or maize
Wheat	1–13	
Sugar (cane/beet)	1–21	One study estimates a figure of 21 per cent. The other studies estimate figures of no more than 2 per cent
Studies that analyse the impacts of global/multi-regional biofuel mandates		
Oilseeds	2–7	
Vegetable oils	35	There is only one study cited which looks at this
Cereals/maize	1–35	
Wheat	1–8	
Sugar (cane/beet)	9.2 or 11.6	Only one study cited looks at this. These are the estimates from the two scenarios modelled

Source: Kretschmer *et al.* (2012: 48).

shock to corn prices will be mitigated in the long run, by global supply responses through competitive international markets (this is a theme that recurs throughout much of the literature). They thus conclude there is 'no long-run food before fuel issue' (Qiu *et al.*, 2012: 2021). Bastianin *et al.* (2013) look specifically at Nebraska, the second largest ethanol-producing state in the US. They also find no evidence for the food v. fuel debate, given the absence of a significant impact of ethanol prices on the price dynamics of agricultural commodities.[7]

When Du and McPhail (2012) examined these relationships, they found that, prior to March 2008, their results were very similar to those reported above, in particular that the prices of ethanol, petrol and corn were found to be affected most by own-demand shocks. That said, from about March 2008 a strong and two-way link emerged between corn and ethanol – a price change for one had a significant impact on the variance in price of the other (very roughly 25 per cent in each direction; Du and McPhail, 2012: 186). This they attribute to policy developments around that time, as more corn went for ethanol and more ethanol was blended into petrol. Du *et al.* (2011), find similar results when looking at crude oil, corn and wheat markets. Elmarzougui and Larue (2013) locate the structural break earlier, at around 1999, but agree that policy has led to stronger cross-market linkages over time. The results of Nazlioglu *et al.* (2013) are also broadly similar. The one commodity, post-crisis, they find still not to be linked strongly with oil prices is sugar, a finding broadly consistent with the study by Serra (2011).

The results of Gardebroek and Hernandez (2013) fall in between the preceding studies. On the one hand, like Du and McPhail, they find a structural break – in their case in 2006 – after which the links between corn and ethanol are stronger than previously. On the other hand, they find volatility spillovers go from corn to ethanol, but *not* from ethanol to corn, a finding more in keeping with studies cited earlier. They also find no evidence of *major* spillovers from oil to corn markets, although such spillovers as did occur did get stronger after 2007.

McPhail *et al.* (2012) also find mixed results, in their study of the possible impact on corn prices of global corn demand, energy (both oil prices and ethanol production) and commodity speculation (the buying and selling of commodity futures primarily for financial gain). This latter variable has received increased attention since the 2007–2008 food price spike (we report further on this below). McPhail *et al.* find that whilst speculation has a significant short run effect, in the long run it is energy that impacts significantly on corn prices.[8] They also identify a complicating factor that, as we shall see later, arises in many studies. The link from oil to corn can be direct (in the sense of not passing through ethanol markets) via, for example, higher fertiliser and other input costs in corn production, transport costs and so on, but it can also be indirect, via ethanol. Higher petrol prices increase the demand for ethanol, which can increase the price of corn as more is demanded for the production of ethanol. Overall, they find no clear impact on corn prices, of corn being demanded for ethanol.

A multi-country study which also splits its analysis up into distinct time periods is that by Ciaian and Kancs (2011). They also find that links from energy to agricultural commodities are growing over time. Moreover, these links are stronger with biofuel production than without. Thus they also find that the indirect links from higher oil prices to commodity prices via biofuels are stronger than the direct links from higher oil prices to agricultural commodity prices. The fact that the findings from this broader study are consistent with the previous, narrower, studies has a further important implication: studies with a narrower focus can still give us valuable insights into the broader biofuels picture.

As noted above, speculation is one of the factors seen as contributing to recent commodity price movements. But whilst futures markets can be a space for speculation, they also handle a lot of 'traditional' activity, such as hedging against adverse price movements. Natanelov *et al.* (2013) utilise the information provided by futures market data to analyse a range of interlinkages, similar to those investigated by McPhail *et al.* They establish significant links from crude oil to corn markets and from crude oil to ethanol markets. They also find the links between ethanol and corn are more complex. As with several other studies, they identify an important role for US policy in driving ethanol production, and, in line with some of the studies cited above, growing links from ethanol to corn.

They go further, however, and identify a complex mix of factors that 'results in a scenario of various forces pulling the markets in different directions in function of present-day market conditions' (Natanelov *et al.*, 2013: 511). Furthermore, as noted with previous studies, higher oil prices make ethanol more attractive, increasing the demand for corn for ethanol. But in their study, Natanelov *et al.* find a threshold effect. Rather than there being a smooth relationship, there exists a key value – crude oil passing $75 per barrel – that marks a shift in the ethanol-corn relationship. Given that the Brent spot price (annual data) has been above this level more or less continuously since the 2007–2008 price spike, this finding offers a particular reading of why ethanol-corn price links have been stronger in recent years.

Another study to utilise futures market data, but in a broader context than the US, is Algieri (2014). In keeping with other studies, she finds commodity prices to be influenced by a complex set of drivers. She also finds that energy market returns have a significant impact on commodity returns. Using the Standard & Poor's 500 Index,[9] Algieri finds significant impacts of stock returns on sugar, wheat and soybean oil. In addition, oil and ethanol returns impact on sugar, wheat, soybeans and corn: energy markets increase commodity price volatility. She argues that biofuels policies should be carefully monitored, to ensure they do not create food v. fuel conflicts, but she does not argue that her results indicate such a situation exists already.

The foregoing studies provide a number of features to highlight. The first is the extent to which their findings vary – in these cases, by producing potentially diametrically opposite findings, for example, that US ethanol production either does, or does not, have a significant impact on corn prices. Several find that links between oil, biofuel, and agricultural commodity markets grow stronger in recent years, as policy pushes production higher. Most studies, however, also state that there is no simple relationship between prices in different markets and that other factors than biofuels are also at work. Moreover, several studies make the point that liberalised commodity markets can help demand and supply responses accommodate and counter any policy-induced price impact from biofuels.

We now turn to consider one particular aspect of this in more detail – the extent to which biofuels may have played a role in the recent food price spikes, in 2007–2008 and 2011. Interviews in Washington DC revealed great frustration amongst biofuels interests at the approach of some elements of the anti-biofuels lobby. They were incredibly active and vocal in 2008, as commodity and food prices rose, and as ethanol production rose, but went very quiet when, in 2009, ethanol production continued to rise, but commodity prices fell, and by rather more than grocery prices in the shops (several studies cited in this chapter make particular reference to the Grocery Manufacturers Association, who helped coordinate anti-biofuels lobbying in 2007–2008). Once again we ask what the studies have found that have investigated this in detail. Again, we consider a range of studies which are representative of the studies undertaken and of estimates produced.

One feature of these studies, just as with those looking at general linkages and transmission channels, is their acknowledgement of multiple factors working together to drive prices. As a result, they all acknowledge that biofuels probably played a role – but, as previously, the challenge is to isolate the magnitude of that contribution. It should also be noted that some of the studies we look at were published soon after the 2007–2008 price spike. In 2009, prices fell again, challenging some of the arguments presented in those initial studies. The 2011 price spike thus offers something of a natural experiment, by which we can see if analysis of the 2007–2008 spike still held after the price fall in 2009 and spike in 2011. We start with Table 9.2, which summarises some of the estimates of the impact of biofuels on agricultural commodity prices. This is a useful starting point because it includes studies which also include the period prior to the 2007–2008 spike, and those which project forward for the forthcoming decade.

To this we can add a few estimates, presented by Mueller *et al.* (2011: 1629), which refer to biofuels contribution to the 2007–2008 price spike. An estimate from the IMF put the figure at 20–30 per cent. One study looking just at the US put the estimate in the range 6–11 per cent, for both corn ethanol and soybean biodiesel. A second, looking only at corn ethanol, put the figure at 3 per cent, including indirect and spillover

Table 9.2   The contribution of biofuels policies to food prices and to recent food price spikes

Author	Contribution of biofuels to the price rise
Mitchell	66 per cent over 2002–08
Rosegrant	30 per cent, over 2000–07
Wright	Substantial price effect due to biofuels
USDA	13–18 per cent over 2007–08
Taheripour	9–16 per cent over 2001–06
FAO	7–15 per cent over 2008–18
OECD	5–16 per cent over 2008–18
Banse	7–12 per cent over 2008–20
EU Commission	3–6 per cent to 2020 (cereals only)
Von Witzke	0.1–4.6 per cent over 2007–08
Gilbert	Hardly any effect by biofuels
Baffes and Haniotis	Hardly any effect by biofuels
Tangermann	10–30 per cent over 2006–08

Source: Schmitz (2012: 22). Available at: http://www.biokraftstoffverband. de/index.php/hunger.html?file=tl_files/download/Stellungnahmen_und_ Studien/12-02-23-Ufop%20VDB-Schmitz_Vorstudie_ENG.pdf.

effects between crops. Moreover, this study attributed 18 per cent of the price spike to demand factors in emerging economies.

Clearly there is variation in the range of the estimates produced. Schmitz (2012: 21–24) analyses possible reasons why. Longer timeframes of analysis are expected to see smaller price effects, because there is scope for supply (and non-biofuels demand factors) to respond to an initial price rise. Second – and consistent with so many other studies – Schmitz points out that price effects are the result of the complex interactions of multiple factors, making the contribution of any one factor difficult to isolate. Indeed, Schmitz argues that no model embraces all potentially relevant factors. He also makes an extremely important point, drawing on the analysis of Babcock (2011), but which also refers back to a number of the studies discussed earlier: to what extent is a biofuel effect caused by biofuels policy?

Mandates may drive a certain volume of biofuel demand, but if a rise in oil prices raises demand for a biofuel to substitute for it, that is a response driven by market forces, not the mandate. That said, of course, the presence of a mandate can still add to these market effects. Thus Schmitz, citing Babcock, indicates that in the US from 2004–2009, the impact of ethanol policy on the rise in corn prices is estimated at 13 per cent, but the market influence (higher crude prices driving demand for ethanol) is nearly 50 per cent. This leaves nearly half of the rise in corn prices explained by the many other factors identified previously. For wheat, the respective figures are 5 per cent, nearly 30 per cent and nearly 70 per cent.

One of the studies referred to by both Mueller *et al.* and Schmitz is Mitchell (2008). This warrants special attention for two reasons – it produced by far the highest estimate of the impact of biofuels on food prices in the 2000s, and second, because it was widely quoted, typically without any attempt at qualification or clarification. To correct this oversight, the relevant text is worth quoting in full (Mitchell, 2008: 16–17). Following a summary of a range of influencing factors, Mitchell then writes:

> Thus, the combination of higher energy prices and related increases in fertilizer prices and transport costs, and dollar weakness caused food prices to rise by about 35–40 percentage points from January 2002 until June 2008. These factors explain 25–30 percent of the total price increase, and most of the remaining 70–75 per cent increase in food commodities prices was due to biofuels and the related consequences of low grain stocks, large land use shifts, speculative activity and export bans. It is difficult, if not impossible, to compare these estimates with estimates from other studies because of different methodologies, widely different time periods considered, different prices compared, and different food products examined, however most other studies have also recognized biofuels production as a major factor driving food prices.

The immediate problem is that it is not clear from this just how much of the 70–75 per cent is actually caused by biofuels, given that Mitchell lumps biofuels in with a range of other factors. Speculation in commodity markets, already discussed, was influenced, for example, by the possibility for higher returns from primary commodities during the Great Moderation.[10] On the question of stockholding policy FAO (2009), for example, shows that stocks have been declining since the 1990s, whilst several studies pick up on more recent changes in Chinese stockholding policy. Thus the challenge is, rather, to estimate the role biofuels policies have played in contributing to these trends over this period.

The statement regarding land use is worth particular attention, and we analyse this in detail below. More broadly, land-use shifts have occurred – but again, how much of those changes are attributable to biofuels? Data from FAOSTAT show wheat area declining throughout the

1980s and 1990s, rising again only from about 2004 (yet throughout this time, rising yields ensured production continued to rise). Meanwhile, interviews in Washington confirm that at least some of the expanded corn production went onto land previously used for soybeans, but that some soybean production moved onto land previously used for cotton, the area of which was falling for entirely unrelated reasons. Thus the only commodity which, in this little sequence of events, saw a categorical net fall in area was the one non-food crop: cotton.

As for export bans or restrictions, these were put in place by several countries, often as a result of domestic food security and food price concerns, which followed extreme climatic events that reduced domestic production (see also, for example, Dollive, 2008) – not biofuels. The impact on world prices was especially dramatic as some of these countries would normally be exporting to the world market. Yet the only reference Mitchell (2008: 16) makes to climate relates to Australia, where back-to-back droughts reduced exports by about 4 per cent 'and other exporters would normally have been able to offset this loss', making no further reference to the situation in other countries.[11]

Again, this is not to deny that biofuels had any role in the price spike, but the magnitude of the estimate by Mitchell does not really stand up to scrutiny, at least insofar as it has been interpreted as indicating 70–75 per cent of the price spike can plausibly be attributed to biofuels. This brings us to the last part of the quote. As we have seen, different studies do indeed come up with different estimates – but no other study produced an estimate anywhere near as large as that by Mitchell. This is not *proof* of error or of different methodology, but if Mitchell's estimate is to be taken as broadly accurate, the obvious next question is why so many studies would produce estimates that were similarly, consistently, inaccurate or otherwise so different? Mueller *et al.* (2011: 1631) are rather more forthright in their assessment of Mitchell's study, when they argue that 'the method used in the Mitchell report appears to be inappropriate and overly simplistic'. They go instead with the other studies, which produce an estimated range for the contribution of biofuels to the 2007–2008 price spike of 3–30 per cent – and as Table 9.2 shows, even the 30 per cent figure is an outlier.

Following Mitchell (2008), it is also instructive to look at Baffes and Dennis (2013), a more recent study, published after the 2007–2008 and 2011 price spikes, and also conducted by World Bank researchers. They note that whilst the price spikes appeared initially to be similar to those experienced in the 1950s and 1970s, they now appear to be different insofar as, after a period of declining real prices, real price rises 'have a more permanent character' (Baffes and Dennis, 2013: 2). They focus on five commodities, four of which are part of the biofuel story (corn, wheat, soybeans and palm oil), one of which is not (rice). They find more than half of the rise in prices is attributed to crude oil prices (with figures of 52 per cent for corn, and 64 per cent for wheat), with stocks (stock to usage ratios) and exchange rates each contributing about 15 per cent on average (although it contributes 22 per cent to corn stocks, with the exchange rate having no effect on corn prices). The model of rice prices performs least well, which they attribute to rice markets being subject to the greatest policy distortions (the exchange rate contributed 29 per cent, which indicates a major role of domestic policy in influencing domestic prices for rice as a staple).

When comparing their results with those of other studies, Baffes and Dennis (2013: 15) argue that the limited impact of stock/usage ratios is in line with other studies, as is the finding of the importance of crude oil prices. Quoting von Witzke and Noleppa (2011), they also indicate that rising crude prices affected freight rates. Furthermore, without giving detailed figures, Baffes and Dennis (2013: 15) argue that the effect of biofuels 'on food prices is not as strong as has been reported in previous studies' (for example, Mitchell, 2008). That said, one recurring theme in this chapter is that crude oil prices affect biofuels, through both direct and indirect channels. On this point, Baffes and Dennis (2013: 13–14) are quite clear. Given the importance of this, we quote them at length (emphasis added)[12]:

To the extent that production of biofuels is driven by mandates, the mixed evidence on the link between energy and food prices should not be surprising. To see this, consider an exogenous shock which pushes crude oil prices up, in turn, lowering fuel consumption. Under a mandated ethanol/gasoline mixture ethanol and maize prices will decline, *ceteris paribus*, leading to a negative relationship between food and oil prices (de Gorter and Just 2009). From a statistical perspective, the mixed evidence on the energy/non-energy price linkage during the recent boom may reflect the frequency of the data used in various models. Indeed, Zilberman *et al.* (2012) noted that higher frequency (and hence 'noisier') data are typically associated with weaker price relationships. The key conclusion from the studies based on time series analysis – especially the ones that include the recent boom period – is that *the strength of the energy/food price relationship should not be used as a metric associated with the impact of biofuels on food prices*.

To this, we can identify other qualifications to estimates of the impact of biofuels on food prices. von Witzke and Noleppa (2011: 7–8) point out that expansion in the production of biofuels over the last ten years has involved just 2 per cent of global cropland (and if we allow for even a small amount of land-use substitution between different agricultural activities, biofuels have used an even smaller net percentage change in total agricultural land). Moreover, they make a point made by others, that because 'the growth in bioenergy production has been fairly continuous, it is not likely that bioenergy has contributed much to the commodity price spike', a point which has particular significance in that, whilst price levels may be higher, than recently the price spikes, by definition, have included significant price falls as well.

The significance of the impact of biofuels on volatility is questioned in many studies. On price levels, we should also note the point made by the OECD (2008a) that, the point about total land use notwithstanding, the use of agricultural commodities for biofuels represents the largest source of new demand to have come into commodity markets in many years (OECD, 2008a). This will impact on price formation through both the demand and supply sides, but is likely to impact more on general price levels than the recent spikes. In the longer run, as already noted, changes in prices can lead to re-equilibrating changes in both supply and demand. There is also the longer-term question of whether biofuels policies are likely to continue to drive up demand for agricultural commodities. We return to this below.

To add to this analysis, Trostle *et al.* (2011: 7) present data from four international commodity price indices, all adjusted to a base of 2002. All indices show a clear rise from 2002 – a period before the US introduced the RFS (2005), and well before the EU introduced its mandatory blend requirement in the RED and QFD of 2009. Trostle *et al.* also make the point that the supply and demand factors driving the price rise over 2002–2006 contributed to the spikes in 2007–2008 and 2010–2011. In addition, there were short-term factors driving the two recent price spikes, including sharp changes in demand with economic crisis and recovery. They also refer to changes in stocks, attributed to weather-related swings in supply.[13] Again, this is indicating multiple factors at work.

These studies indicate biofuels represent a new source of market demand, which could affect price levels. Huchet-Bourdon (2011), meanwhile, looks specifically at the issue of price volatility. Looking over half a century of data, she finds no evidence that volatility has increased systematically over this timeframe: the rate of volatility is, itself, volatile.

Recent spikes follow a broadly similar pattern to previous examples – a sharp rise, then fall, in aggregate *food* commodity prices, accompanied by a general rise in commodity prices, notably crude oil and metals (something that numerous studies have confirmed for the recent spikes).

She finds volatility in recent years was higher than in the 1990s, but not higher than the 1970s, other than for wheat and rice. As already indicated, rice is subject to particular policy interventions, whilst numerous studies have shown that wheat prices have been susceptible to extreme weather events. For beef and sugar, recent volatility was lower than the 1970s whilst, for other commodities, there was no statistically significant difference between the two periods. With sugar, this is broadly consistent with the studies cited earlier. Sugar (and beef) is also the least correlated with the crude oil price. When looking at individual years, however, for many crops other than sugar, 2008 was a year of particularly high volatility, although only dairy products and soybean oil saw this higher volatility continue into 2009.

Recent volatility in agricultural commodity markets is, in general, no higher than in the 1970s. Huchet-Bourdon is careful to point out, however, that with international trade being more liberalised, such volatility as does occur may well now get transmitted more quickly. Moreover, volatility remains a policy concern, and cautions that the increased price link with crude oil for some commodities during the 2000s 'may confirm' a role for biofuels in recent price spikes (Huchet-Bourdon, 2011: 28). Liapis (2012) asks if higher volatility is a result of thinner commodity markets (lower exports as a percentage of total production). Using three different measures, he tests this hypothesis and finds that this is not the case. Indeed, this finding is consistent with other studies, insofar as trade liberalisation would, other things being equal, lead to thicker (Liapis talks of deep or liquid) markets, with a higher percentage of commodities produced subsequently being traded (see also Abbott *et al.*, 2011).

An issue noted by Huchet-Bourdon, and explored in detail by Ott (2014), is the difference in volatility estimates, depending on whether one is looking at higher frequency (for example, monthly) data, or annual data. In recent years, volatility has increased notably for inter-year price volatility. Looking at intra-year volatility, the main driver is a low stock to usage ratio, whilst factors such as the exchange rate and macroeconomic conditions play only a minor role. In contrast, longer-run conditional volatility on returns is driven mainly by competition and trade volumes. International trade reduces inter-year volatility, confirming that policies such as export bans have a significant impact on volatility. Stocks play a modest role, whilst trade conditions play a more important role, as do crude oil volatility and the exchange rate (confirming the results of several earlier studies). Interestingly, Ott finds evidence that speculation activity can reduce price volatility, in the short term (reflected in the intra-year volatility measure) and the longer-term (reflected in conditional volatility).

But are we fussing over nothing? Does food price volatility even matter? Barrett and Bellemare (2011) argue that price levels are the primary source of problems, not price volatility, but accept that confusion is understandable, as the two variables are correlated. Price levels, they point out, are high but, making the same point as Huchet-Bourdon (2011), volatility is not particularly out of line historically, and is lower than in the 1970s. To illustrate their point, Barrett and Bellemare contrast stylised facts about consumers and producers, especially in poor countries. They argue that net consumers of food are typically poorer than producers. The former group are affected adversely by high food prices, whereas the latter group can benefit from them. In contrast, they suggest that volatility does not necessarily affect consumers – who can substitute different foodstuffs, given also that price volatilities for different foodstuffs tend to be less than perfectly correlated.[14]

Producers, however, are affected by volatility: first, they must commit investments in seeds and other inputs before knowing what output (and revenue) will be, and second, in facing such uncertainty, they may decide to underinvest, reducing plantings and thus potential output. As stylised facts, these arguments hold together, but as Roache (2010) finds, lower frequency volatility shows a positive correlation across different commodities. This would, as a minimum, limit consumers' ability to substitute between foodstuffs. Thus volatility may affect not only producers, but also consumers – and to a greater degree than Barrett and Bellemare suggest.

The work of Barrett and Bellemare does, however, raise an extremely important point – different individuals or groups within society may benefit or suffer from higher prices, and may suffer or benefit from lower prices. Swinnen (2011) explores this by asking a deceptively simple question: what is the right price of food? He does not offer a definitive answer. Instead, he bases his analysis around the fundamental paradox inherent in the recent debate, presents an eloquent critique of how this paradox is communicated, and points out the challenges to policy-making that arise from the failure to diagnose clearly the different dimensions of the paradox. The central issue is very simple – low food prices harm food producers but benefit consumers, whilst high prices harm consumers but benefit producers. Failure to recognise the winners as well as the losers in each situation can lead to ineffective policies.

Swinnen provides quotes which, as recently as 2005, were identifying only those harmed by low prices, followed in more recent years by quotes focusing only on those harmed by the more recent high prices. The punchline is that it was the same organisations being quoted in each case. Examples are given from publications by NGOs (Oxfam and the Bread for the World Institute), but Swinnen argues that this is understandable as they are advocacy groups, using information selectively to push their agenda. What Swinnen also shows is that international organisations such as the FAO, IFPRI, OECD, World Bank and the IMF have produced reports, before and after the recent price spikes, which also show that same selective analysis. Unrelated to Swinnen's analysis, but related to his point about the existence of upsides as well as downsides in policy, interviewees in the US noted that if biofuels keep corn prices above a certain level, this avoids triggering federal payments under the farm programmes, thus saving the government money – which, in the presence of the fiscal cliff, could be seen as a very positive side effect.

So where does this leave us? There are studies which failed to find a statistically significant link from biofuels to commodity markets. There is one study which attributed up to three quarters of the 2007–2008 price spike to biofuels. In between there are numerous studies which argue that biofuels have impacted on price levels and price volatility, but the quantification of those impacts indicates biofuels has, in all likelihood, had a relatively modest effect – and certainly more modest than some of the more extreme rhetoric would suggest. Even then, it is not clear whether, or how much, of this is driven by mandates or by market forces.

Moreover, whilst the establishment of mandates has been driving demand levels higher, it is not clear for how much longer this will continue. In the US, the component of the RFS for which corn ethanol is eligible is very close to exhaustion. In the EU, the mandate allows for substitution between feedstocks; in total represents a somewhat smaller volume of biofuels than the US; and faces the prospect of having a cap imposed on the contribution to the overall mandate that first generation biofuels can make. The foregoing analysis has also indicated that a distinction can be drawn between short run impacts of higher demand on markets, and the longer run when supply is able to respond. Much is made in the foregoing studies about the pass-through of prices and volatility from oil markets and prices, to commodity markets, both

directly and indirectly, but some of these effects will be driven by market forces rather than mandates. There is analysis of different types
of volatility, the different causes, and the different consequences. And
what we have also seen, is that when looking at the impact of changes
in price levels, we need to remember that with both price rises and price
falls, there are individuals and groups within society that will benefit as
well as be harmed.

The challenges posed by indirect land-use change (ILUC)

We now turn to Indirect Land-Use Change, another policy challenge
which arises from the implementation of biofuels policies, but one
which policy-makers are attempting to address at the policy design
stage. ILUC, along with the more general Land-Use Change (LUC), and
sibling concept Direct Land-Use Change (DLUC) were introduced and
discussed in Chapter 1. We now return to ILUC, to analyse it in more
detail. The title of this section deliberately has a double meaning: ILUC
presents a potential challenge to biofuels policies and policy-makers,
insofar as it could affect the total emissions attributable to a given
volume of biofuels. The marketability of those biofuels would thus be
challenged, under policies which define eligible biofuels in terms of
total attributed emissions. Second, given that ILUC cannot actually be
observed directly, plus the number of other factors that could reason-
ably explain land-use changes, ILUC presents a considerable challenge
to anybody seeking to estimate its existence and magnitude.

In this section, we review some of the vast literature estimating ILUC
effects, but we also look at the work being undertaken that is starting
specifically to examine the analytical challenges presented by ILUC.
Indeed, we start with the latter body of work, to provide context and
caveat to the subsequent review of the empirical literature. We thus
focus primarily on the challenges to estimating ILUC, rather than to
the estimates that have been made. As important as the techniques
and data demands are for modelling ILUC, the issues are so numerous
that we cannot provide comprehensive coverage. Further references are
provided as we go through, for the interested reader to follow up with.

The challenges to estimating ILUC effects

There is an emerging consensus that points to improvements over time
in our understanding of, and ability to model, ILUC, but that limitations
remain and uncertainties are still many (see, inter alia, Overmars *et al.*,

2011; Brunelle and Dumas, 2012). These uncertainties lead Dumortier *et al.* (2011: 428) to argue that, since ILUC lies at the heart of policy debate, 'given the available knowledge, it is very difficult to narrow the range of reasonable parameter values to tighten the set of results to a level that would allow robust policy conclusions'. Di Lucia *et al.* (2012: 12) point out that if policy-makers respond to uncertainty by taking action anyway, but based on science that is uncertain, they may do so by framing the uncertain science as actually providing certainty, thus creating the so-called 'uncertainty paradox'.

Di Lucia *et al.* (2012: 17, emphasis added) connect the issues of uncertainty and policy advice to argue that scientists should refrain 'from claiming the need for more research, better models and more comprehensive data, and admit that due to *irreducible* uncertainties there are many valid results, but no ultimate correct ones. In this way, they will not give policy-makers the opportunity to misuse the scientific knowledge'. Moreover, 'policy makers could solve the current policy deadlock by taking an openly political decision on ILUC', using 'the available scientific knowledge ... as a tool to support, monitor and assess policies that have been selected through the political process ... [rather than] ... as a predictive oracle to guide policy choices'. Or, because we shall never be able to model ILUC precisely, policy-makers should make decisions which recognise the scientific work on ILUC, but which are ultimately a political calculus into which those studies are but one input.

We do not comment on the merits of this assertion, but we believe it is reasonable to suggest that the EU approach to ILUC, in part, reflects this advice. The proposal of adding fixed values to emissions calculations for different first generation biofuels feedstocks, as discussed in Chapter 3, recognises that something should be done (in keeping with the general tenor of discussions within the EU), recognises that different feedstocks, on average, have different ILUC effects, but does not get bogged down in trying to ascertain exact ILUC estimates for each feedstock.

As a result, however, the EU proposal to add ILUC factors to first generation biofuels of 12 grams of CO_2-equivalent per megajoule of biofuel energy for cereals, 13 for sugars and 55 for (all) oil crops is a blunt and unnuanced response. For example, the report by Copenhagen Economics (2011), published prior to the EUs formal reform proposals, considers a range of policy options available to the EU. It provides evidence that suggests palm oil is a biodiesel feedstock which has the potential to deliver significant GHG emissions savings (alongside, notably, sugarcane ethanol). That said, the EU proposal also includes the ongoing monitoring of scientific developments, giving scope for ILUC factors to be changed in the future, if deemed appropriate.

Witcover *et al.* (2013) offer suggestions as to how to deal with LUC which, again, broadly approximate to the EU approach, and also with a suggestion that more nuance is needed. They argue the adoption of feedstock-specific ILUC factors 'is an effective way to signal acceptable LUC risk' (Witcover *et al.*, 2013: 71), selected in terms of values which lie within the ranges of estimates determined scientifically. They also argue, however, that this is not sufficient – efforts should also focus upon promoting those feedstocks which have lower or no land-use impact; promoting investment in efforts which increase land productivity (see below for further consideration of this aspect of biofuels feedstock production); and promoting measures protecting the carbon-storage and environmental services of land.

But what does it mean for estimates to lie within a range that has been determined scientifically? Plevin *et al.* (2010: 8015) argue that ILUC effects are uncertain, but may be much greater than estimated. They reach this conclusion by comparing default ILUC values from CARB and the EPA (30g CO_2 per megajoule and 34g CO_2 equivalent per megajoule, respectively) with their own model-derived ranges of emissions from possible ILUC effects: a bounded range of 10–340 g CO_2 per megajoule and a 95 per cent central band of 21–142 g CO_2 equivalent per megajoule. We do not wish to debate these estimates. Rather, we use them to highlight the point that, regardless of the modelling uncertainty, the estimation of a *range* of values does not necessarily tell us where within that range the 'true' value lies. Clearly the CARB and EPA figures lie very much towards the lower end of the range of estimates from Plevin *et al.*, but they also lie within the 95 per cent bounds. One of our interviewees in Brazil expressed great frustration that the debate over biofuels seemed to be based on a starting point of 'worst case scenarios' – something, he argued, that simply is not done with anything else.

Another aspect of the politics which, given the arguments in Part I of this book, is seen to be very important, is made by Overmars *et al.* (2011: 256). Biofuels policies stand on three legs, of which climate change mitigation is but one. They agree that ILUC effects 'could' (Overmars *et al.*, 2011: 248) raise emissions above the levels generated by fossil fuels, but they also point out that because there are other factors to consider when assessing biofuels policies, policy-makers need to make a judgement about how to weigh each of the three policy drivers against each other.

A particular problem in quantitative work is confusing correlation and causality. Two data series, A and B, may move together, but that may be because A drives B; B drives A; something else, C, drives A and B; or maybe C drives A; something else again, D drives B; and C and D are causally related. Or it may just be coincidence: the fact that a statistically significant correlation has been found between the number of nesting storks in Sweden and the Swedish birth rate does not mean that more babies require more storks to deliver them to their parents.[15] Kim and Dale (2011) sought to use historical data to estimate ILUC and thus avoid the modelling challenges outlined above. Whilst appealing in theory, it triggered a response (O'Hare *et al.*, 2011) that tore their approach apart. They argued Kim and Dale muddled correlation and causality, made completely incorrect inferences from their statistics, and also misunderstood ILUC 'which is not whether biofuels production causes total deforestation to increase, but whether it causes it to increase more, or reforestation to increase less, than would happen without the biofuel program' (O'Hare *et al.*, 2011: 4486). For the record, Kim and Dale could find no evidence that US corn ethanol production triggered ILUC, but that was not what concerned O'Hare *et al.*

Whilst much of the research on ILUC focuses on mandate-driven EU or US policy, Andrade de Sá *et al.* (2013) look at links between the expansion of Brazilian sugarcane production in São Paulo State and deforestation and cattle ranching in the Amazon. They confirm the expansion of sugarcane area in São Paulo State and an increase in cattle numbers in the 'legal Amazon'. Bearing in mind the criticisms of O'Hare *et al.* aimed at the work of Kim and Dale, it is interesting to note the following quote from Andrade de Sá *et al.* (2013: 387): 'it is not possible with our data to establish spatial causality between sugarcane expansion in the state of São Paulo and deforestation in the Amazon. Yet, our study is the first to provide empirical evidence that such a relationship between these two land uses might exist'.

Andrade de Sá *et al.*, therefore, appear to be claiming both that causality cannot be established from their dataset, but also that they are the first to show that such a relationship (by which, from this statement, they must mean causality) might exist. Moreover, they 'interpret as displacement' (Andrade de Sá *et al.*, 2013: 386) the observed spatial substitution of land use – which presumably also implies causality. They then argue that this substitution can take 10–15 years, without explaining where the cattle might have gone in the interim. Brazil's cattle population grew from 147.1 million head in 1990 to 211.3 million head in 2010, which indicates another possible explanation for cattle ranching expanding into previously unfarmed areas.

We can also triangulate the results of Andrade de Sá *et al.* with our own data from interviews with senior government officials in Brasilia. One point made to us was that average stocking density is rising amongst Brazil's cattle herds, something confirmed by data in Andrade de Sá *et al.* The latter argue that it remains lower in the legal Amazon than in São Paulo State, which is correct, but it is rising sharply in both regions. Something else that was raised in our interviews in Brasilia was the impact of land price differentials. On average, we were told that land on the frontier was an eighth to a tenth of the price for prime farmland – and could be up to one-thirtieth, which made such land a very attractive investment proposition. Furthermore, we were informed that following the deforestation of such land, cattle will sometimes be brought onto the cleared land for a few years, to earn the landowner an income before the land is put to other uses, given that it can take five years or more for the roots of the biomass to die away and leave the soil suitable for crop-based agriculture.

We are not suggesting that Andrade de Sá *et al.* are wrong in their core estimations. Rather, they are being rather bold in their interpretation of those estimations – in particular in terms of seeing correlations as possible evidence of a causal link. Our interviews in Brasilia, even if taken as representing an 'official' government view, nonetheless identify factors that plausibly suggest caution is required when looking for evidence of a causal displacement relationship from sugarcane expansion in São Paulo State to cattle ranching on deforested land, a thousand or more kilometres and 10–15 years away.

Estimating ILUC

One observation from the literature discussed above is that, related to the problem of ILUC modelling per se, cross-study comparisons are risky, because different economic models use different techniques[16] and make different assumptions, all of which can explain at least some the differences in estimated ILUC effects. Indeed, the report from Copenhagen Economics (2011: 5) suggests that even when similar countries, feedstocks, etc. are being modelled, the resulting estimates can have up to an eleven-times difference.

Perhaps the best place to start is with Searchinger *et al.* (2008). This was the study that really kick-started the debate over ILUC, even though the concept was known about before then. The headline figure is that the emissions from corn ethanol could be nearly double the fossil fuel replaced, when land-use change is included. The authors therefore argued that focusing on producing biofuels from waste products is preferable to using land-based feedstocks. Regardless of the actual emissions estimated, this study showed clearly that ILUC is an issue that warrants attention. But the study triggered responses that were concerned about that estimate.

Very soon after the publication of Searchinger *et al.* (2008), a letter was published in *Science* that was co-authored by the developer of the GREET economic model Searchinger *et al.* used.[17] The most significant issue is that Searchinger *et al.* modelled the impact of doubling corn ethanol production above a benchmark level of estimated production in 2016. The problem is, the volume of corn ethanol modelled is almost exactly twice the level permitted under the RFS. Thus the results of Searchinger *et al.*, paradoxically, confirm the merits of the cap imposed in the RFS, as discussed in Chapter 4. Furthermore, the letter argues that there was, at the time, no evidence of ILUC occurring in other countries, because US corn exports had held steady, whilst US exports of biofuels by-products that could be turned into animal feed, had risen. This letter contains further points of detail that debate various values used in the Searchinger *et al.* paper, all of which serve to demonstrate how tricky modelling ILUC actually is.

In the same volume of *Science* as Searchinger *et al.* (2008) was an article by Fargione *et al.*, another study to receive much attention. They state that 'converting rainforests, peatlands, savannas, or grasslands to produce food crop-based biofuels in Brazil, Southest Asia and the US creates a "biofuels carbon debt" by releasing 17 to 420 times more CO_2 that the annual greenhouse gas (GHG) reductions that these biofuels would provide by displacing fossil fuels' (Fargione *et al.*, 2008: 1235). This is, unambiguously, an important point but it confirms something already in several key policies – the importance of establishing and enforcing sustainability criteria, to ensure these land-types are not used to grow biofuels feedstocks. Regarding alternative feedstocks, Fargione *et al.* agree with Searchinger *et al.* about promoting biofuels from waste, but they also estimate cellulosic ethanol will deliver lower emissions than fossil fuels. Thus neither Fargione *et al.*, nor Searchinger *et al.*, are against biofuels per se, just against biofuels done badly.

In a similar vein, Melillo *et al.* (2009) argue that ILUC can raise significantly the emissions from biofuels production, and they argue that avoiding deforestation will help avoid ILUC effects. In addition, they point to the need to control nitrogen fertiliser use as they argue that, potentially, nitrous oxide emissions could have a greater warming effect than carbon. Havlík *et al.* (2011: 5699) offer similar policy advice about sourcing, in particular, second generation feedstocks, but supplement this with the conclusion that 'we recommend policy action to focus directly on the positive and negative environmental and social effects linked with biofuel production, rather than on biofuel production itself'. This points back to the notion that biofuels are not intrinsically good or bad, just done well or done badly.

Since 2008, a mini-industry has sprung up devoted to modelling ILUC. Rather than try to summarise this literature by identifying ILUC estimates and drawing comparisons between studies, we instead direct readers to some of the papers which do this job for us. Fortunately for the non-academic reader, several of these are published online in sources that are freely available, in contrast to subscription-based academic journals.

An early study which informed EU debate on ILUC is Edwards *et al.* (2010). This compares several studies and considers how ILUC could be handled in EU policy. One dimension of this study is that there appears to be a downward trend in estimated ILUC effects, if the models are arranged chronologically. That said, the study itself does not draw any inferences from this. To do so would be to risk inferring causality from what may be correlation (or mere coincidence). This does, however, highlight an issue that warrants monitoring – does our growing understanding of ILUC and capacity to model actually have a systematic effect on the estimates produced? Even if it does, the caveat of uncertainty remaining over ILUC still applies.

Several of the studies analysed above base their analysis on a comparison of studies, a meta-analysis.[18] But what might the new features be that are being incorporated into newer studies? To illustrate this, we consider Goodwin *et al.* (2012), who estimate the price elasticity of crop yields, and Langeveld *et al.* (2013), who analyse the impacts of multiple cropping.

Goodwin *et al.* investigate how US farmers respond to price changes, in terms of how yields vary. The traditional approach to ILUC would see a causal chain from US farmers selling more corn for ethanol production, to the world corn price rising, to corn producers in other countries changing their production decisions as a result. Goodwin *et al.* find that US corn producers, in addition to having a long-run yield response to price changes of 0.25 (consistent with other studies), also respond to price changes *within a given crop year*, if the price comes early enough in the cycle: 'at the mean values, if the percentage rate of change in prices rose from its average value of 1% to 2%, yields would increase by about 0.1%' (Goodwin *et al.*, 2012: 13). Significantly, this quantitative estimate broadly confirms discussions with farmers in focus groups. What this tells us is that if a price rise occurs (maybe as a result of corn going for ethanol production), and if it occurs early enough in the season, US corn farmers will respond, through crop management, to get more corn from each hectare of land. Thus a rise in the US corn price, which may plausibly be influenced by a biofuel end-use, could trigger a rise in US corn production, which could mitigate any knock-on effect on world corn prices, and thus lessen the (indirect) impact of US corn producers' land-use decisions on corn producers elsewhere.

Langeveld *et al.* (2013) look at the issue of multiple cropping practices, where more than one crop is grown on a given area of land in a given year. Their study also includes a further dimension of the production of (some) biofuels: by-products that are sold for animal feed – important given that much of the corn crop goes for animal feed (for example, DDGS from corn). Biofuels-derived animal feed can substitute for feed based on inputs grown specifically for that purpose. This, in turn, means that the total land area required to produce a given amount of feedstocks for biofuels and for animal feed is reduced. This can be very significant, given that 'A third of the grain that goes into ethanol production comes out as DDGS'.[19]

Because these by-products increase total production from a given area of land, this is equivalent to a de facto increase in the actual area of land used. This is also how Langeveld *et al.* treat multiple cropping in their model. They argue that whilst, between 2000 and 2010, land used for producing biofuel feedstocks rose by 14 million hectares, increased cropping intensity generated, in effect, over 42 million hectares of additional crop land. They thus conclude that 'biofuels *cannot* be identified as the most important or single global cause of land-use change. Other drivers have caused more (and more permanent) loss of agricultural area including process of urbanization, infrastructure development, tourism and even conversion into nature ... observed changes in land use caused by biofuel policies are very small in comparison to other changes' (Langeveld *et al.*, 2013: 57).

An alternative to increasing the number of crops grown on land each year, is improving the agronomic practices to deliver higher yields for the crops that are grown. Looking at 20 key feedstocks for ethanol and biodiesel, Johnston *et al.* (2011) estimate that bringing yields up to median levels could deliver feedstocks sufficient to produce 112.5 billion litres of ethanol and 8.5 billion litres of biodiesel. They recognise that agricultural intensification can, in turn, cause environmental problems, although basing estimates on lifting yields to the median, rather than any higher, may help mitigate those consequences (they also estimate yields rising to the 75th and 90th percentiles, but do not report those results in the paper).

Regarding the production of biofuels by-products, we also mention Taheripour *et al.* (2010), as representative of this literature.[20] They analyse DDGS and vegetable oil by-products and argue that studies which ignore biofuel by-products overstate the impact of first generation biofuels on global agricultural and land markets. With their inclusion, price effects are curtailed and ILUC effects moderated. Wallington *et al.* (2012) offer a similarly cautionary tale, based on their analysis which incorporates by-products and intensification of land use through rising yields (akin to the focus of Goodwin *et al.* rather than the multiple-cropping of Langeveld *et al.*). Indeed, these mitigating effects on prices and ILUC could be reinforced if farmers pursued both yield increases and multiple-cropping.

It should be made clear that these recent studies no more prove the insignificance of ILUC, than earlier studies proved its significance. They do, however, demonstrate clearly our evolving understanding of the factors that need to be incorporated into any estimate of ILUC; they reinforce just how complex are the economic interactions that determine the presence, magnitude and possible location of ILUC effects; and they therefore reinforce the need for caution when reading any given study on ILUC. Goodwin *et al.*, to take a very good example, try to improve the robustness of their findings by combining quantitative and qualitative (focus-group) methods of analysis. But even this cannot predict whether farmers might behave differently in ten or twenty years' time. Equally, technological and agronomic advances may result in even higher price-yield responses. What is found for US corn producers may not be found in, for example, German oilseed rape producers, or Malaysian palm oil producers.

Ultimately, therefore, ILUC is a concept that is very hard to conceptualise formally and to model. Indeed, as quoted earlier, Di Lucia *et al.* (2012) argue that some of the uncertainties surrounding ILUC are irreducible. Thus our knowledge may grow, but it will converge on total understanding only asymptotically. As a result, policy-makers face difficult choices: to pursue developments in understanding and modelling ability; and if so, how long to wait until they feel knowledge has reached a point where a policy can be put in place that avoids the aforementioned 'uncertainty paradox'; or to make no allowance at all for ILUC; or to do something that is primarily politically based, which recognises the possibility of ILUC, but which is modest enough not to undermine the biofuel industry; or to pursue the precautionary principle and limit the scale of the (principally first generation) biofuels industry.

Or, is a totally different approach again needed? Analysis by Palmer (2014) indicates that much of the discourse around ILUC has been controlled by the Commission, in such a way that (as indeed we have done in this section), the focus has been on technical emissions issues, abstracted from a sense of place. Moreover, it has avoided consideration of the 'many profoundly moral and ethical dimensions of the issue' (Palmer, 2014: 349).[21] A detailed analysis of these dimensions is beyond the scope of this book, but we highlight this as an area where a broader debate is warranted: are biofuels policies essentially technical exercises in addressing, for example, emissions concerns; or are they, should they be seen as, more than this? We also look at some of these issues below, in considering public perceptions and media representations of biofuels policies.

A running theme throughout this book has been the inability of policy-makers and industry to traverse the 'technology bridge' and develop significant volumes of advanced biofuels, although as Searchinger *et al.* warned several years ago, some land-based non-food feedstocks can also trigger LUC effects. As noted in Chapter 5, several interviewees spoke of policy getting ahead of the science, something that refers, in general, to the failure to cross the technology bridge and, more specifically, to the challenges of the ILUC debate. This debate, because of its inherent uncertainties, will always be a focus for discussion and disagreement. The challenge for policy-makers is whether even to support the development of our understanding of ILUC (given some of the arguments, above, about its irreducible uncertainties), whilst continuing to support the development of biofuels which can minimise or even avoid ILUC effects altogether (see also Lankoski and Ollikainen, 2011, for a broader perspective on these challenges). The question is – for how long can policy-makers withstand the pressures the ILUC debate brings, supporting the continuation of a first generation biofuels-based industry whilst waiting for the right sort of advanced biofuels to come to market? And is this sufficient to address the moral and ethical dimensions highlighted by Palmer (2014)?

Consumers, consumer engagement and the acceptability of biofuels

And so, nearly at the end of the book, we finally bring us, the end-users of biofuels, into the debate. This reflects the extent to which consumers featured, during our extensive fieldwork for this project, as part of the policy-making process. Throughout this project, we have had a strong sense that, in the EU and US in particular, the underlying principle was that so long as biofuels were delivered to market, and so long as fuel companies blended biofuels into existing fossil fuels, then that was all that mattered.[22]

As we have seen throughout this book, however, biofuels intersect many other issues that are of great concern to consumers. These include fuel prices, food prices, the environment, deforestation and land use. As a result, engaging consumers directly with biofuels can be a key piece of the policy jigsaw, especially in terms of boosting market growth through the greater consumer acceptability of and support for biofuels. As it is, the focus of most attention (as this chapter attests) is on potential downsides of biofuels. Indeed, as Swinnen (2011) demonstrates, even organisations from which we might expect objective analyses are delivering rather one-sided messages. It is perhaps telling that, in this debate, academics have also come to the issue of consumer engagement and involvement only recently. Compared with development of the biofuels themselves, 'social acceptance issues have been underestimated' (Chin *et al.*, 2014).[23]

Much of the work we focus on here addresses, broadly speaking, two related issues. One asks consumers directly what they think about biofuels. This involves either survey techniques or, in at least one case below, focus groups – where participants are brought together and their views on a topic ascertained via what is, in effect, an extended group interview. The second set of research studies analyses how different issues are presented in the mass media. These two issues are interlinked as the mass media (rather than official government publications of scientific publications/academic research) will be citizens' main source of information on a topic. One inevitable consequence of the research methods employed is that most of this research is location-specific, looking at citizens and media in a single country (or, in one case, comparing two small countries), or even (in some US studies) looking at the views of citizens in a single state.

We first consider studies analysing citizens' views in the US, carried out in Indiana and Wisconsin, states where ethanol is a significant part of the economy.[24] Despite this, Delshad *et al.* (2010) found that whilst citizens were knowledgeable about biofuel technologies, they were much less so about biofuel policies. The view they did have on policy, however, was that they liked least the policy option actually in place at the time: the fixed subsidy – the 45 cent VEETC (see Chapter 4). There was limited support for corn ethanol, a slight majority in favour of biofuels overall, and stronger support still for second generation ethanol. Cacciatore *et al.* (2012a) find that, amongst more knowledgeable participants, the perception of benefits from biofuels relative to risks was lower. This study, looking at 'biofuels', is broadly consistent with that of Delshad *et al.*, where there was little support for corn ethanol, the dominant and most high-profile biofuel in the US.

Delshad *et al.* found that, in their study, citizens' beliefs about the economic and environmental impacts of biofuels were more important than factors such as energy independence. Cacciatore *et al.*, meanwhile, found support for biofuels was greater amongst younger participants, and with females over males. Given the widely established finding in the literature that females are more risk averse than males when it comes to new technologies, Cacciatore *et al.* speculate that this concern may be dominated by another factor also identified in wider research, that females are more pro-environment. This, the authors emphasise, is currently only speculation as to why they found females more supportive of biofuels than males. Cacciatore *et al.* also find Democrats are more likely to support biofuels than Republicans, although they do not explore whether this might, in part, be explained by the gender gap in the US voting patterns, where there is greater support for Democrat presidential candidates amongst female than male voters.

In a study we look further at below, Cacciatore *et al.* (2012b: 679) find that males are more likely than females to see biofuels as leading to higher food prices. They speculate that since females are the principal shoppers in most homes, those who do the shopping are not actually witnessing a noticeable impact of biofuels on prices, whereas those who, on average, do not do the shopping believe otherwise. Why they might believe otherwise is something we consider later, when we look at the role of the media in public perceptions about biofuels.

The complexity of the issues related to biofuels comes out strongly in the work of Jensen and Andersen (2013), in Denmark. They indicate that, given concerns over climate change and the environment, the initial perception of biofuels was positive. This was especially so regarding second generation biofuels and the use of waste as a feedstock (Jensen and Andersen, 2013: 46). As possible problems with biofuels were introduced into the discussion, however, people did start to alter their views. Rather than go from broadly supporting, to broadly opposing biofuels, citizens incorporated the complexity of the issue in their own, increasingly nuanced, views. Jensen and Andersen (2013: 55) thus conclude that, in Denmark, there is 'ambivalent and conditional acceptance' of biofuels, where 'the contingencies have a number of different components'. There is, therefore, no simple for or against view about biofuels.

Returning to Cacciatore *et al.* (2012b), this study also explores citizens' opinions – but does so with a twist. Drawing on ideas that we shall explore further below, it looks at whether using the word 'ethanol' or 'biofuels' affects people's responses in answering a survey. The short answer is yes, it does. Specifically – and in keeping with some of the findings already reported – the word ethanol is received less positively than biofuels. Furthermore, whilst this has little impact on Republicans (who are more opposed to biofuels, regardless of wording), Democrats are strongly influenced by this wording. The authors speculate that one reason for the more negative opinion is that it is closely aligned with 'corn ethanol', which brings (possible) food price effects to mind. Given the 'rural development' driver of biofuels policy identified in Chapter 1, it is interesting that rural residents were more likely than urban residents to agree that biofuels brought jobs. Cacciatore *et al.* conclude from their findings regarding the importance of how the product is referred to, that more research is needed on the media representation of biofuels issues – which brings us to the second theme identified in the research literature, the mass media.

One of the challenges with research involving survey or interview methodology is, because it is typically a snapshot taken at a certain point in time, determining the direction of causality between variables from the results can be difficult or impossible. That said, there is a growing body of literature that is showing evidence of a link between citizens' perceptions of biofuels and media coverage, even if we cannot always determine whether they are linked causally and, if so, which is driving which. But, as Cacciatore *et al.* (2012b) demonstrated, simply by changing one word in their survey and giving the two different versions to broadly matching samples, the presentation of ideas matters.

Delshad and Raymond (2013) analyse, first, how newspapers framed the biofuels issue over the period 1999–2008. They find an increasingly negative representation of the issues over time – with particular attention paid to the possible negative impacts on consumers. The authors then follow this up with a survey, conducted in 2010, to see how much these media reports affect perceptions, in relation to other factors such as political affiliation or identification, local economic interests and whether an individual self-classifies as an environmentalist. In general they find that new media do influence public opinion. Digging deeper, they found (perhaps surprisingly) that people from Midwestern states, whilst more supportive of biofuels, were not more supportive than others of biofuels policies. This may align with the earlier findings of Delshad *et al.* (2010) regarding people's differing knowledge of biofuel technologies and policies – especially if they are then getting information about biofuels from a media which is presenting biofuels increasingly negatively. Delshad and Raymond do, however, find greater support for second and third generation biofuels across the sample as a whole.

One curious finding is that individuals who identify themselves as environmentalists are more likely to be in favour of biofuels (including corn ethanol, so this is not a labelling issue). Although these authors also find Democrats more pro-biofuels, whether this outcome might be explained by environmentalists being more likely to be Democrat – and that effect dominating – is not explored. One finding of particular interest in the context of our own research is that Delshad and Raymond found biofuels to be a much more partisan issue than they had expected – indeed, a more important factor in determining attitudes towards biofuels than any other. For reasons we explore further below, this stands in contrast to the findings of the interviews we conducted in Washington DC amongst policy elites.

Another point Delshad and Raymond make on this partisan question is that this split is found to be stronger for biofuels than for other energy questions. It is important to note that biofuels do sit in this broader policy context, as Part I demonstrated in all three of our focus cases. Wolbring and Noga (2013) look at this broader context and find that the media are extremely important in shaping public opinion over greening and energy issues more generally. They investigate the reporting of greening and energy in four Canadian newspapers, two national, two regional. Whilst 88 per cent of articles on greening also mention energy, only 0.15 per cent of articles on energy also mention greening. Thus whilst energy is a greening issue, greening is not seen as much of an energy issue. This, they argue, will make it harder for citizens to become more fully informed about, and engaged with, the greening of energy.

A recent European study addresses similar issues, but in the context of a comparison of Norway and Sweden (Skjølsvold, 2012). His focus is 'bioenergy', including biofuels. This study explores how technology is 'domesticated' in different ways by the media. Skjølsvold expected, ex ante, that the focus of the media would be on issues of controversy – but found this was not the case. What he found was that, in Norway, the presentation was more ambivalent (interestingly, the same word as used by Jensen and Andersen to describe Danish attitudes). In Sweden, however, the approach and the tone were much more optimistic. Skjølsvold (2013: 526) finds that Norwegian representations were centred around economic and technical ambivalence, and possible food v. fuel problems. In Sweden, the focus was more on how technology can help solve problems, plus how bioenergy was related to green consumption – although there was also some recognition of possible food v. fuel concerns.

Skjølsvold (2012: 517) observes that in Norway, energy policy has tended to focus on 'large scale national projects', such as hydroelectricity, oil and gas, and – more recently – carbon capture and storage (CCS). In other words, Norway's approach has been to try to de-carbonise its existing energy matrix. In Sweden however, the approach has been to have bioenergy at the heart of a process of technology replacement – a process that actually began in the 1970s. Thus Sweden has sought to de-carbonise its energy matrix through the use of new technologies to deliver new energy sources. Indeed, in the context of domestic debates over nuclear versus bioenergy-based power, the recent expansion of bioenergy sources 'was a political relief' (Skjølsvold, 2013: 517). Bioenergy has therefore been cast as a source of local and national pride.

An interesting point of comparison with this study comes from an investigation by Sengers *et al.* (2010), who analyse Dutch media coverage of biofuels, combined with interview data from discussions with practitioners. They find both the more positive Swedish-type perspectives, where technology can help deliver solutions to problems, as well as negative views which accord, to a degree, with the Norwegian outlook. Between 2000 and 2008, they identify a trend from more positive perceptions, based on what they refer to as a techno-economic/regional frame, towards much more negative perceptions, driven by those with a social–ecological/global frame. This represents a shift from an emphasis on those who believe technology can deliver solutions in Europe, to those who focus on the potential downsides of biofuels and harm they can cause the environment and people in developing countries. In common with much of the debate around biofuels, they find very little nuance in the coverage.

Given the importance of the media in people's understanding of issues surrounding biofuels, it is clear that the use of language is critical. Ribeiro (2013) takes a broader look at the social aspects of biofuels and, in so doing, makes a point that links back to the ILUC debate. A part of the biofuels debate that we have not explored is where new land can be brought into agricultural production, perhaps specifically for biofuels feedstocks, which does not violate the demands of sustainability criteria. One such type of land is 'marginal land'. The problem with this, as Ribeiro (2013: 360) points out, is that 'The absence of an agreed definition of marginal land should serve as an example of how values that are in dispute can lead to misinterpretation of crucial sustainability aspects'. If neither scientists nor policy-makers can agree on the definitions of key concepts in the biofuels debate (especially when they relate to something as controversial as ILUC), then it is inevitable that the information citizens will be relying on for their understanding will be less than clear.[25]

In all of this discussion of consumer behaviour and preferences, Maréchal (2009: 81) makes a very interesting point: 'domestic energy consumption...is not visible....This implies that people do not consider the remote environmental impacts of their actions when performing energy-related behaviors'. What the foregoing suggests is that, with biofuels, this is not the case. There are different views based on political affiliation, age, gender and so on, but the fact that biofuels even *might* impact on land use, deforestation and other effects that cannot be observed directly as being linked to biofuels, is as relevant as the possible visible impacts, such as petrol prices and food prices. The role of the media in conveying information about both visible and invisible impacts, whether neutrally or with a particular slant, is crucial. Indeed, information generally – or its absence – is very important (Savvanidou *et al.*, 2010).

Thus far we have just been referring to the media, as in 'mass media'. Talamini *et al.* (2012), in their research, identify three different sources of information – journalists (mass media), scientists (research publications) and the government (public policy documents). They analysed, in total, over 3300 documents, published between 1997 and 2006. Their principal findings (Talamini *et al.*, 2012: 654) were that policy-makers have drawn on the work of both journalists (who focused more on the economic, environmental, political and geopolitical dimensions), and on scientists (who focused more on the environmental, agronomic and technological dimensions). They also find a rebalancing over time, whereby the concerns upon which journalists focus are becoming more prevalent in government publications, vis-à-vis scientific outputs.

Perhaps reflecting the timeframe of the data analysed, another of their findings was that the growth in ethanol production preceded the growth of liquid biofuels on the agendas of all three sources of information. There was some interplay between output and government publications, but then as production really started to expand, so in its wake did the attention of journalists, scientists and policy-makers. An important conclusion they draw from their findings is that whilst the policy was, at the time, based more on scientific perspectives, this may change over time. Certainly, if we put this finding in the context of the other research reported above from the US, subsequent developments in US policy (the EISA), very recent developments in alternative domestic energy sources, and in the light of current debates over the RFS mandates, then their sense that the reference to scientists in policy-making, relative to journalists might wane, appears to have foundation.

This last study is important because, unlike the studies that focused on citizens' views, it also incorporates the views of policy-makers, through official documents. Delshad *et al.* (2010) argued in their study that, whilst they focused on citizens, the gap between the policies in place and the policy preferences expressed by people in the focus groups, pointed to the need to look also at the views of policy elites. Meanwhile, Delshad and Raymond, also with references to the views of citizens, found biofuels split less along regional lines and more along partisan party lines than other energy issues.

This repeated finding, of views split along partisan lines amongst citizens, is extremely interesting in the light of our interview data. Among the many people we interviewed in Washington were several who worked either in the relevant Departments and Agencies of government, or who were staffers working for the relevant committees in the Senate and House of Representatives – that is, we interviewed 'policy elites'. These interviews conveyed a broadly consistent picture that ethanol was, in fact, a bipartisan issue (one interviewee described it as 'reliably bipartisan') – where you came from mattered far more than which party you represented. Talamini *et al.* (2012: 654) note that the 'expanding ethanol industry in the US appears to wield its political power over the government as a way to obtain economic incentives for ethanol production. That is, ethanol production appears to be a public policy-derived process'. But what is shaping this political power? One plausible explanation is that it comes from the strength of having regional (corn belt), bipartisan, support.[26]

So what can be done to boost social acceptance? There is no simple answer to this question, but Chin *et al.* (2014) make the point that there is a sequence that should be followed: address the acceptance of issues relating to biofuels feedstocks before addressing the acceptance of biofuels. Given the discussion throughout this chapter, that is sound advice. The use of agricultural feedstocks for first generation biofuels can lead to food v. fuel concerns, which then brings in concerns such as higher food prices. It can also lead to ILUC concerns, which then brings in concerns over the ability of biofuels to deliver lower GHG emissions. As the analysis in this section has shown, these concerns are very real for citizens and have a clear impact on their perceptions of biofuels, regardless of the actual magnitudes of these effects. But, as Chin *et al.* point out, they are located primarily at the start of the biofuels supply chain. Sorting out those issues can help make biofuels more acceptable to consumers at the other end.

They suggest a series of measures to help this process – based primarily on their study of Malaysia, but with elements of general applicability. In particular, they recommend moving towards second generation biofuel production. They predict opposition from existing biofuels producers and investors, but since both US and EU policies limit the scope for first generation biofuels to be used, and since work is ongoing to develop commercial second generation production, this concern may be over-stated. As has been clear throughout, one of the biggest difficulties facing the industry and policy-makers is precisely this development, even though both EU and US policies are predicated on it happening.

As a result, if Chin *et al.* are correct in their analysis, and if we therefore need to focus first on getting consumer acceptability at the feedstock end of the supply chain, the current inability to deliver adequate supplies of second generation biofuels to market means that we either need to curtail production of first generation biofuels, regardless of existing mandates, or there needs to be better communication of the complex issues surrounding biofuels. Mohr and Raman (2013) explore what lessons second generation biofuels can take from first generation biofuels, in terms of a sustainability appraisal. In analysing these possible lessons, they make a point that is pertinent to the current situation: they highlight 'the limitations of focusing on narrow framings or understandings of core sustainability challenges, such as the now ubiquitous "food vs. Fuel" conflict. Thus "food vs. fuel" is a simplification of a complex array of interrelated factors not least to do with how land is valued, managed and governed' (Mohr and Raman, 2013: 121).

We thus finish as we began the book, and have continued throughout this chapter – the issues surrounding biofuels are multidimensional and extremely complex. Arguably, we need to explore each issue separately, in order to make our analysis tractable, but, as this chapter has shown, these sometimes seemingly disparate issues are all linked, in one way or another. Thus recognition is also required of the bigger picture into which biofuels fit.

Conclusions

In this chapter, we have explored in detail some of the particular challenges facing biofuels policy-makers – the food v. fuel debate, ILUC and consumer engagement with biofuels, in particular as mediated through mass media. With both food v. fuel and ILUC, the economics underlying the headline issues are extremely complex, which creates uncertainty over policy design, and ambiguity in policy implementation. This uncertainty manifests itself in the wide range of ways in which economists have modelled the interactions which are might give rise to food v. fuel and ILUC effects.[27] It also creates space for debate and disagreement over the actual impacts of biofuels production and policies. As we have seen, not only have partisan NGOs adopted distinct advocacy positions on biofuels, so too – the evidence suggests – have some of the leading international organisations undertaking analysis of biofuels policies.

We have also seen how mass media will frequently adopt a particular stance on this issue (as others) which, increasingly, is having as significant an impact on people's perceptions of biofuels as more 'traditional' factors, such as prior political beliefs, gender, age and so on. We have seen how even the word used ('ethanol' or 'biofuels') can influence people's perceptions, let alone whether the focus is more on the perceived benefits or the potential downsides of biofuels.

At the start of the book, we identified three drivers of biofuels policy – energy security, rural development and climate change mitigation. These have all been used, at different times and with different emphases, in our three focus cases of Brazil, the EU and the US. They can, at times, be contradictory, and they create opportunities for biofuels to be supported even when, in one dimension or another, those biofuels are not necessarily 'biofuels done well'.

In this chapter we have seen another internal policy contradiction: the food v. fuel debate implies a zero-sum game for food production, whereas ILUC is predicated on new land being brought into food production. Swinnen (2011) shows how actors with a range of vested interests emphasise only the negative dimension of the prevailing low or high commodity and food prices. We argue here that, in a similar way, certain issues surrounding land availability, land use and its place in the food v. fuel and ILUC debates are also carefully exploited. Most notably, by isolating the two issues and debating each one separately, exposure of the inherent contradictions can be avoided.

Whilst biofuels and biofuels policies have extremely important technical dimensions, in this book we have viewed biofuels policies, the challenges they seek to address and the challenges they can give rise to, as an essentially political problem. In particular, they are seeking to reconciled diverse interests and incommensurable values. And whilst there are tricky technical questions in, for example, the debates over food v. fuel and indirect land-use change, these have been used in a partial and tendentious fashion by vested interests, for and against biofuels and, within this, for different types of biofuel and biofuel policy. These politics of biofuels policy endure, and the policy capacity to deal with them remains uncertain. Thus, in turn, so too does the future for biofuels, at least on their current scale and in their current forms.

Notes

1 Biofuels and Biofuels Policies – An Introduction

1. The question of the depletion of finite resources is distinct from the issue of 'peak oil'. This latter term refers to the point where maximum production from reserves is attained, after which it begins to decline. This will be affected by a number of factors, notably the scale of known reserves, and the technical ability to extract marginal volumes of oil from those reserves. Thus whilst oil reserves are, by definition, finite, peak oil is determined by our knowledge about those reserves – their location and scale – and our ability to get the oil out.

2. Oil, as a percentage of total primary energy consumption, fell in Western Europe from nearly 60 per cent in 1973, to 42 per cent in 2009. For Japan, the figures were 78 per cent to 47 per cent (Chalabi, 2010: 189).

3. This phrase was used by several interviewees during the fieldwork stage of our research.

4. This section draws on a range of resources. See, for example, Energy Future Coalition (2007); Rajagopal and Zilberman (2007); FAO (2008); OECD (2008b); IEA Bioenergy (2009); Soetaert and Vandamme (2009); Auld (2013); Brown and Brown (2012).

5. There are a great many sources available which address the science of biofuels. See, inter alia, Soetaert and Vandamme (2009); Langeveld *et al.* (2010). Several of the general references utilised in this section also provide overviews, with varying degrees of technical detail.

6. Some varieties of rape are, especially in the US, known as canola.

7. This distinction in processes is reflected in a distinction sometimes made, between biodiesel and renewable diesel. The former is produced by transesterification and is blended in diesel, typically, at up to 20 per cent. The latter is produced by hydrogenation, and can be used in existing infrastructure and existing engines, in any blend (see US Energy Information Administration note: http://www.eia.gov/todayinenergy/detail.cfm?id=16111).

8. <http://news.mongabay.com/bioenergy/2007/10/quick-look-at-fourth-generation.html>. Last accessed 18 November 2013.

9. No page numbers are given on this website.

10. Data from the Global Renewable Fuels Alliance website (http://globalrfa.org/biofuels-map/). Last accessed 26 February 2014.

11. See, for example, Edwards *et al.* (2008: 14–15).

12. NUTS = Nomenclature des Unités Territoriales Statistiques, ranging from NUTS 0 (whole country), via progressive disaggregation, to NUTS 5 (local municipalities and communes).

13. Our sole focus in this book is biofuels policies. It is beyond this narrow scope to debate the alternative or complementary means of addressing energy security, rural development and climate change mitigation. For a useful introduction to those wider issues, see Charles *et al.* (2007).

14. In one case, Delzeit *et al.*, undated, the authors are clear on page 2 that this change does not constitute DLUC, but in Figure 1 on page 3 appear to change their minds.

2 Brazilian Biofuels Policy – An Introduction and Overview

1. Throughout this chapter, 'ethanol' and 'alcohol' will be used interchangeably, depending on the context of use.
2. Puerto Rico *et al.* (2010: 1876) provide a table summarising the main legislation prior to 1975.
3. Ministério da Agricultura, Pecuária e Abastecimento.
4. http://www.agricultura.gov.br/arq_editor/file/Desenvolvimento_Sustentavel/ Agroenergia/estatisticas/anuario_cana_ingles.pdf
5. Puerto Rico *et al.* (2010: 1884–1886) discuss in detail the role of the state oil company, Petrobras, in Proálcool; see also Hira and Guilherme de Oliveira (2009).
6. Conselho Interministerial do Açúcar e do Álcool.
7. As explained to the authors in an interview with a former senior government official.
8. Conselho Nacional de Politica Energética.
9. Ministério de Minas e Energia.
10. The source data also provides a split between the CS and NE regions of Brazil.
11. Programa Nacional de Produção e Uso de Biodiesel.

3 EU Biofuels Policy – An Introduction and Overview

1. This Treaty also created the European Union. We use the term 'EU' in this book as a shorthand, regardless of the point in time being discussed.
2. For more detail on the emergence and role of the EU as climate policy leader, see Oberthür and Pallemaerts, eds. (2010); Wurzel and Connelly, eds. (2011); Bourblanc (2011).
3. Formally, the United Nations Conference on Environment and Development.
4. Article 8(2)d of Directive 92/81/EEC 'on the harmonization of the structures of excise duties on mineral oils'. Official Journal L316, page 14, 31.10.1992.
5. Official Journal C209, 29.7.1994, page 9.
6. COM(95)682 final, 13.12.1995. See especially page 35.
7. Available at: http://ec.europa.eu/energy/library/e2020fd.pdf
8. COM(97)599 final, 26.11.1997.
9. A commitment made by the EU15, Bulgaria, the Czech Republic, Estonia, Latvia, Liechtenstein, Lithuania, Monaco, Romania, Slovakia, Slovenia and Switzerland.
10. Official Journal L283, 27.10.2001, pp. 33–40.
11. http://www.europarl.europa.eu/summits/lis1_en.htm (last accessed 28 February 2014).
12. Available at: http://ec.europa.eu/governance/impact/background/docs/ goteborg_concl_en.pdf
13. See also Howes (2010), amongst others.

14. Council Directive 2003/96/EC of 27 October 2003 restructuring the Community framework for the taxation of energy products and electricity. Official Journal L283, 31.10.2003, pp. 51–70.
15. Full details of the Climate and Energy Package can be found at: http://ec.europa.eu/clima/policies/package/index_en.htm (last accessed 2 October 2013)
16. European Energy to 2020: A Scenario Approach. Energy in Europe Special Edition, Spring 1996. http://ec.europa.eu/energy/library/e2020fd.pdf
17. http://www.consilium.europa.eu/ueDocs/cms_Data/docs/pressData/en/ec/89013.pdf, page 15.
18. Page 21 of Council document 7224/1/07. Available at: http://www.consilium.europa.eu/ueDocs/cms_Data/docs/pressData/en/ec/93135.pdf.
19. Annex 1 of Council document 7224/1/07. Available at: http://www.consilium.europa.eu/ueDocs/cms_Data/docs/pressData/en/ec/93135.pdf. This came a year after the European Council had called for the creation of an EPE: http://www.consilium.europa.eu/ueDocs/cms_Data/docs/pressData/en/ec/89013.pdf
20. This paragraph represents our interpretation of our interview data – our interviewees did not state this explicitly.
21. Di Lucia and Kronsell (2010) analyse in detail the reasons behind the uneven implementation of the 2003 Biofuels Directive. Based on their analysis of the evidence, these authors classified non-implementing countries as being either willing but unable, unwilling, or unable. An interesting topic for future research would be to see whether those same factors explained the variation witnessed here, bearing in mind that the biofuels targets laid down in the RED and FQD are mandatory.
22. In addition, Regulation 443/2009 requires a reduction in car emissions over time. See: http://eur-lex.europa.eu/LexUriServ/LexUriServ.do?uri=OJ:L:2009:140:0001:0015:EN:PDF
23. Sustainability criteria raise concerns over whether they represent a non-WTO compliant barrier to trade, either intrinsically or in their implementation. We address this issue in Chapter 8.
24. Communication from the Commission on the practical implementation of the EU biofuels and bioliquids sustainability scheme and on counting rules for biofuels. *Official Journal of the European Union*, C160, 19 June 2010, pp. 8–16.
25. See also Commission Decision of 10 June 2010 on guidelines for the calculation of land carbon stocks for the purpose of Annex V to Directive 2009/28/EC. *Official Journal of the European Union*, L151, pp. 19–41.
26. See the opinion piece by Eric Sievers, CEO of Ethanol Europe Renewables Ltd on the Euractiv website on 3 March 2014: http://www.euractiv.com/energy/better-biofuels-policy-analysis-533863 (accessed 17 March 2014). See also Oliver (2013).

4 US Biofuels Policy – An Introduction and Overview

1. US Senate and House of Representatives (2005).
2. US Senate and House of Representatives (2007).

3. For a detailed analysis of the links between OPEC and the US energy and security policies, see Chalabi (2010).
4. Erickson *et al*. (2014) analyse the revised EPA approach, as well as estimate the possible emissions consequences of reduced annual biofuels mandates under the EPA's new methodology.
5. These data, and the rest of this discussion, are taken primarily from 'Obama Messes with the RFS', published online on Biofuels Digest on 17 November 2013 and accessed on 18 November 2013: http://www.biofuelsdigest.com/bdigest/2013/11/17/obama-messes-with-the-rfs/
6. Brent Erickson is Executive VP and head of the Industrial and Environmental section at the Biotechnology Industry Organization. This quote comes from an article entitled 'EPS's Christmas Gift to the Oil Refiners', Biofuel Digest, 29 December 2013: http://www.biofuelsdigest.com/bdigest/2013/12/29/epas-christmas-gift-to-the-oil-refiners/_ (accessed 6 January 2014).
7. http://www.biofuelsdigest.com/bdigest/2013/12/04/stand-by-me-renewable-fuels-defenders-pull-out-the-stops-to-persuade-epa-to-continue-the-war-on-imported-oil/_ (accessed 5 December 2013).

5   Comparing Biofuels Policy Drivers – Common Themes, Differences and Issues for Analysis

1. Gee and McMeekin (2011) offer a comparison of US and Brazilian policy developments, starting from the 1970s. In particular, emphasising the role of government policies, they look at how Brazil's industry developed faster than that of the US initially, for the US to catch up and, in terms of ethanol production, surpass Brazil 30 years later. See also Harvey and McMeekin (2010) for a comparative analysis of the EU, US and Brazil.
2. When, in 2011, the authors presented a paper at The World Bank in Washington, one discussant suggested that, at that point in time, the global figure was that 99 per cent of all biofuels were first generation.
3. European Commission (2014: 32, paragraphs 112 and 113).
4. The Seven Sisters were the seven major international oil companies who gained control of the oil concessions in the Middle East from the 1920s. The US companies amongst the seven were Standard Oil of New Jersey (now Exxon), Mobil, Gulf, Chevron and Texaco. For more on the Seven Sisters and how they controlled the oil market and oil prices, see Chalabi (2010), in particular Chapters 1 and 2, *passim*.
5. The speech is available at: http://www.presidency.ucsb.edu/ws/?pid=4051. Accessed 20 January 2014.
6. This was also seen in reverse when, following the collapse in the price of oil in 1986, both the US and UK lobbied Saudi Arabia in particular to help support a higher price of oil, necessary to sustain profitability in the Texan and North Sea oil fields.
7. http://www.americanvaluesnetwork.org/climate/. Accessed 20 January 2014.
8. http://www.rurdev.usda.gov/RD_EconomicDevelopment.html. Accessed 20 January 2014.

6 The Challenge of Policy Capacity in Biofuels Policy Design

1. This chapter draws extensively on, and develops further, the analysis in Kay and Ackrill (2012).
2. Policy conversion is not the same as 'greenwashing'. This latter term refers to misleading claims being made about the environmental benefits, qualities or performance of goods, or of companies. In contrast, as noted earlier, Brazilian policy-makers have responded to discussions over concerns such as ILUC by re-analysing the data and confirming that their sugarcane ethanol really does perform very well against this and other 'green' benchmarks.
3. Note also a stronger link in the EU between the concept of 'rural development' and the realities of smaller farmers, rather than the biggest producers.
4. As seen in Chapter 4, Box 4.1, US policy also embeds GHG emissions reductions thresholds into sustainability criteria applicable to all feedstock production, from any source country. That said, in the US case, the text distinguishes between, for example, federal and non-federal lands, suggesting a primarily domestic orientation dominating their initial design.
5. In interviews, this was always pronounced 'caff-ay', hence the accent.
6. http://ec.europa.eu/energy/strategies/2011/roadmap_2050_en.htm (accessed 5 July 2011).

7 Biofuels Policy Design and External Implementation Challenges

1. This chapter draws on, and develops, a paper presented to the 7th ECPR General Conference, Sciences Po, Bordeaux 4–7 September 2013: Policy Layering as Insulation: Implementing US And EU Biofuels Policy.
2. The rise of historical institutionalism has been a salient trend in the social sciences over the last 20 years. Institutions are defined conventionally as sets of regularised practices with a rule-like quality that structure the behaviour of political and economic actors in policy-making and which cannot be changed easily or instantaneously. Critical historical institutionalist insights have recently been adapted for the more fine-grained concept of policy, by recognising that clusters of governmental decisions, actions and norms can – over time – form policy systems, reinforced by feedback mechanisms, which function as institutions. The crux of historical institutionalism for policy studies is the claim that institutions matter in the analysis of policy change, providing constraints on as well as opportunities for change, and that they emerge and develop within a wide variety of historical processes and sequences.
3. The discussion in the next few paragraphs introduces issues analysed in greater depth in Chapter 8.
4. União da Indústria de Cana-de-Açúcar, the Brazilian Sugarcane Industry Association.
5. https://www.gov.uk/government/uploads/system/uploads/attachment_data/file/255318/rtfo-2013-14-year-6-report-1.pdf

6. A view expressed by some participants at a seminar given by the authors at the World Bank in September 2011. See also Erixon (2009).
7. As discussed at a Roundtable on the Development and Impact of Biofuel Policies, Centre for European Studies, Australian National University, 3 August 2012 (held under Chatham House Rules).
8. This was undertaken as part of the 2010 Quadrennial Diplomacy and Development Review. Details are available at: http://www.state.gov/s/dmr/qddr/.
9. Detailed information on the provisions, including changes from the 2008 Farm Bill, can be found at: http://www.kilpatricktownsend.com/en/Knowledge_Center/Alerts_and_Podcasts/Legal_Alerts/2014/02/The_New_Farm_Act_2014.aspx (accessed 2 April 2014).

8 External Dimensions to Biofuels Policies

1. This chapter, in part, utilises the work published in Ackrill and Kay (2011b).
2. Indeed, the EU is itself a manifestation of these pressures and trends.
3. For more on the place of agriculture in the GATT and WTO, see Kay and Ackrill (2009).
4. GATT 1947 is available at: http://www.wto.org/english/docs_e/legal_e/gatt47_01_e.htm. Full details of the additional content of GATT 1994 is available via: http://www.wto.org/english/docs_e/legal_e/06-gatt.pdf.
5. http://www.wto.org/english/tratop_e/dispu_e/cases_e/ds135_e.htm
6. see, for example, Howse *et al.* (2006); Condon (2009); Harmer (2009); Le Roy *et al.* (2011); Swinbank (2009); Switzer and McMahon (2010).
7. Charnowitz *et al.* (2008); Cheyne (2009); Switzer and McMahon (2010).
8. http://www.wto.org/english/tratop_e/envir_e/labelling_e.htm
9. Moïsé and Steenblik (2011), provide an excellent review of trade measures and PPMs.
10. http://www.wto.org/english/thewto_e/whatis_e/tif_e/agrm8_e.htm#subsidies
11. http://www.wto.org/english/tratop_e/agric_e/ag_intro01_intro_e.htm
12. For example, of the 300 or so disputes brought to the GATT in its lifetime, 40% were related to agriculture (Kay and Ackrill, 2009: 484).
13. For completeness, the Red Box was proposed by the US at the outset of the Uruguay Round negotiations. It would have included measures such as the price support instruments which dominated EU policy – with the goal of removing completely such policy instruments within ten years (the so-called 'zero option'). Utterly unacceptable to the EU, this element of the negotiations was soon abandoned. When the EU reformed the Common Agricultural Policy in 1992, new payments for farmers were introduced which substituted partially for price support, but which did not conform with the Green Box. Strictly, they should have been placed in the Amber Box and cut by 20 per cent. Also unacceptable to the EU the compromise solution, negotiated with the US (who had similar types of payment), was to place those policy instruments in a newly created Blue Box. As Aart de Zeeuw, Chair of the GATT Negotiating Group on Agriculture in the early part of the

Uruguay Round, explained it to one of the authors in a conversation, the Blue Box was then effectively placed inside the Green Box – to keep the two sets of policy instruments separate, whilst protecting the Blue Box from the AMS cuts imposed on the Amber Box.

14. To get a flavour of these debates, from a variety of both policy and legal perspectives, see Howse *et al.* (2006); de Vera (2008); Tarasofsky (2008); Condon (2009); Harmer (2009); Josling *et al.* (2010). Deshpande (2006) looks at how biofuels might feature in the Doha Round of WTO negotiations. See also Switzer (2007); Echols (2009); Swinbank (2009a); Mitchell and Tran (2010); Switzer and McMahon (2010).

15. Further reading on this specific topic includes, for example, Charnovitz *et al.* (2008); Vis *et al.* (2008); Lendle and Schaus (2010); Le Roy *et al.* (2011); Lydgate (2012a; 2012b).

16. WTO Committee on Technical Barriers to Trade Notification G/TBT/N/EEC/ 200. Available at: http://docsonline.wto.org/DDFDocuments/t/G/Tbtn08/ EEC200.doc (last accessed 11 February 2011).

17. G/TBT/N/EEC/200/Add.1. Available at: http://docsonline.wto.org/ DDFDocuments/t/G/Tbtn08/EEC200A1.doc (last accessed 11 February 2011).

18. Official Journal C160 of 19 June 2010, pp. 8–16. Available at: http://eur-lex. europa.eu/LexUriServ/LexUriServ.do?uri=OJ:C:2010:160:0008:0016:EN:PDF (last accessed 11 February 2011).

19. http://europa.eu/legislation_summaries/glossary/open_method_ coordination_en.htm

20. http://ec.europa.eu/energy/renewables/biofuels/sustainability_schemes_en. htm. Details of all schemes cited here are available via this link.

21. http://www.bonsucro.com/

22. http://www.rspo.org/

23. http://www.responsiblesoy.org/index.php?lang=en

24. http://rsb.org/. Until 2013, the RSB focused just on biofuels and was known as the Roundtable on Sustainable Biofuels.

25. Obey the law; respect human rights and labour standards; manage input, production and processing efficiencies to enhance sustainability; actively manage biodiversity and ecosystem services; and continuously improve key areas of the business.

26. At the time or writing, this article was still in press. Thus the final pagination of the published version had not been determined. This is the page number of the In-Press version.

27. http://www.dw.de/fossil-fuel-subsidies-outstrip-renewables-funding-by- billions/a-17465775 (accessed 12 March 2014).

28. See https://www.g20.org/ for details.

29. http://www.wto.org/english/tratop_e/dispu_e/cases_e/ds291_e.htm

30. http://bch.cbd.int/protocol/

31. http://www.un.org/documents/ga/conf151/aconf15126-1annex1.htm

32. This issue, and its possible implications, are analysed further in a number of the references cited earlier, for example, Howse *et al.* (2006); Condon (2009); Harmer (2009); Le Roy *et al.* (2009); Swinbank (2009); Switzer and McMahon (2010). See also Wind (2010); Lamers *et al.* (2011).

33. http://www.arb.ca.gov/fuels/lcfs/lcfs.htm

34. In March 2014, research was presented by CARB which re-estimates these ILUC factors and which raises the possibility of these ILUC factors being adjusted (downwards). We discuss ILUC in more detail in Chapter 9.
35. http://www.ethanolrfa.org/exchange/entry/the-ethanol-shuffle/
36. This and related trade issues are analysed in Meyer *et al.* (2013). The photograph on the front page of their report, of two ships passing each other going in opposite directions, sums up this trade pattern very neatly. See also Yano *et al.* (2012).
37. As we were preparing the final manuscript, an equally bizarre example appeared that related to biodiesel. In this case, it was reported that most Canadian biodiesel is exported to the US, to benefit from the blender's tax credit, with the Canadian mandate satisfied by ... imports from the US: http://www.producer.com/2014/04/canadian-biodiesel-exports-put-mandate-into-question/ (accessed 7 May 2014).
38. http://www.wto.org/english/tratop_e/dispu_e/cases_e/ds267_e.htm
39. https://www.wto.org/english/tratop_e/dispu_e/cases_e/ds266_e.htm
40. http://www.reuters.com/article/2011/01/10/us-ethanol-brazil-mccain-idUSTRE7095P420110110
41. http://www.ustr.gov/trade-topics/trade-development/preference-programs/caribbean-basin-initiative-cbi
42. Farinelli *et al.* (2009), in their study of the factors influencing the demand for Brazilian ethanol exports, are unable to determine the impact of the VEETC. Their general finding, however, is that policy-driven blend mandates are an important determinant of Brazil's ethanol exports.
43. See, for example, http://www.jamaicaobserver.com/business/Jamaica-Ethanol-shuts-down-plant–cuts-31-jobs_6748609
44. Jacoby and Meunier (2010) identify five mechanisms by which the EU seeks to manage globalisation – all of which are evident in the case of biofuels: expanding policy scope; exercising regulatory influence; empowering international institutions; enlarging the territorial sphere of EU influence; and redistributing the costs of globalisation.

9   Biofuels Policy Challenges

1. Hereafter, we shall generally refer to 'food' for simplicity, referring to specific 'feed' uses if appropriate.
2. See also the paper by Thompson (2010), one of very few we have seen which recognises this contradiction.
3. http://www.un.org/apps/news/story.asp?NewsID=24434&#.UygGTrFFBkc
4. Throughout this research project, despite the number of times we have seen the first quote, we have not come across any source that looks at what else he said in that same speech.
5. http://www.imeche.org/docs/default-source/reports/Global_Food_Report.pdf?sfvrsn=0
6. In this chapter, we focus primarily on academic research papers. A number of books on biofuels and food-related issues also have sections or chapters on this issue. See, for example, OECD (2008b); Rosillo-Calle and Johnson (2010); Brown (2012); Brown and Brown (2012).

7. Informa Economics (2011) also reports findings that do not support a strict food v. fuel effect, automatically driving up consumer food prices. This study was conducted for the Renewable Fuels Foundation, part of the US Renewable Fuels Association. Had we started with this it might have been easier to dismiss these findings, given the sponsor. This raises a general point, which relates to both pro and anti-biofuels interests in this polarised debate: if possible, results should be triangulated across different studies, especially where studies may be conducted or sponsored by organisations with a particular interest in the outcome.

8. Du *et al.* (2011) find speculation plays a significant role, but they do not make this distinction between short and long run effects.

9. This is based on the market capitalisation of 500 large companies, listed on either the New York Stock Exchange or NASDAQ. The breadth of companies offers an economy-wide picture of economic performance.

10. The period prior to the economic crisis, characterised by less volatile business cycles, with economic growth, low inflation, low interest rates and easy credit. In these conditions, the pursuit for higher returns led fund managers into higher-risk, more volatile markets, such as those of primary commodities.

11. Mitchell (2008) triggered several rejoinders which offer more detail than we can include here, such as: http://www.abengoabioenergy.com/export/sites/abg_bioenergy/resources/pdf/acerca_de/informacion_tecnica/en/Work_Bank_Critique_Julio08.pdf; http://www.democraticunderground.com/discuss/duboard.php?az=view_all&address=115x163126.

12. In this quote, Baffes and Dennis date the paper by Zilberman *et al.* to 2012, but in their list of References they give the date as 2013. In our references, we follow the latter: Zilberman *et al.* (2013). To clarify, the paper was presented at a conference in 2012, was published online in 2012 and published in the *American Journal of Agricultural Economics* in January 2013.

13. Trostle *et al.* provide a wealth of information that can inform discussion on this matter.

14. This may help explain the use of export restrictions on rice if, in those countries, it is felt that there are few substitutes available for consumers. Barrett and Bellemare also look at possible links between volatility and food riots, seen in several countries recently. They argue that causality typically goes from riots to volatility, not the other way around. It is beyond the scope of this book to look at this issue in detail, but we observe that Berazneva and Lee (2013) analyse the causes of food riots in Africa in 2007–2008 without a single reference to 'biofuel', whilst Pangea (2012: 28) find that 'there is very little room to blame biofuels'.

15. Stewart, 1979: 199–200.

16. for example, computable general equilibrium models, partial equilibrium models, sector-specific models and several others – for an excellent review of the range of modelling possibilities, see CBES (2009).

17. http://www.transportation.anl.gov/pdfs/letter_to_science_anldoe_03_14_08.pdf (accessed 12 February 2014). GREET stands for Greenhouse gases, Regulated Emissions, and Energy use in Transportation.

18. See, for example, CBES (2009); Dumortier *et al.* (2011); Brunelle and Dumas (2012); Di Lucia *et al.*(2012). See also IEA Bioenergy (2011); Broch *et al.* (2013).

19. http://www.ksgrains.com/ethanol/ddgs.html (accessed 13 February 2014).
20. For a detailed and extensive analysis of biofuel by-products, see FAO (2012).
21. Gillon (2014) in the same volume, also looks critically at this fundamentally technocratic approach to nature, analysing the US biofuels policy in this light.
22. Or, to repeat the famous misquotation from the film Field of Dreams, 'build it and they will come'.
23. At the time of writing, this article is available on the journal website only as an 'Article in Press' and thus full citation details, notably page numbers, have yet to be finalised. This quote is taken from the Abstract.
24. As of end-2013, Indiana had the sixth largest operating ethanol production in the US, Wisconsin the eighth largest: http://www.neo.ne.gov/statshtml/121.htm.
25. Shortall (2013) explores definitions of marginal land and some of the assumptions they are based on.
26. As noted earlier, we have also undertaken research analysing the EU sugar policy reforms. Sugar policy (in the US perhaps even more so than the EU) tends to be more resistant to reform than other agricultural commodities – and one argument often put forward is precisely this one: that a strong regional coalition is formed from having a geographically concentrated interest group, including producers of the agricultural commodity (in the EU and much of the US, sugarbeet) and, close by, the processing facilities, creating a high value added industry in areas where other such jobs were few and far between.
27. Zhang *et al.* (2013) provide a systematic analysis of the differences in modelling techniques, assumptions and results from these models.

Bibliography

Abbott, P.C., Hurt, C. and Tyner, W.E. (2011) *What's Driving Food Prices in 2011?* Food Foundation Issue Report. Available at: http://www.farmfoundation.org/news/articlefiles/1742-FoodPrices_web.pdf

Abreu, F.R.e, de Souza Viera, J.N. and Ramos, S.Y. (2007) National Program for the Production and Use of Biodiesel: Guidelines, challenges and prospects. *Revista de Política Agrícola*, Year XVI (an English-language version of volume 2006/3), 5–18.

Ackrill, R.W. (2000) *The Common Agricultural Policy*. Sheffield: Sheffield Academic Press, for the University Association for Contemporary European Studies.

Ackrill, R. and Kay, A. (2009) Historical Learning in the Design of WTO Rules: The EC sugar case. *The World Economy*, 32(5), 754–771.

Ackrill, R. and Kay, A. (2011a) Multiple Streams in EU Policy-Making: The case of the 2005 sugar reform. *Journal of European Public Policy*, 18(1), 72–89.

Ackrill, R. and Kay, A. (2011b) EU Biofuels Sustainability Standards and Certification Systems – How to seek WTO-compatibility. *Journal of Agricultural Economics*, 62(3), 551–564. An early version of this research is available freely at: http://www.ntu.ac.uk/research/document_uploads/102325.pdf

Ackrill, R. and Kay, A. (2012) Sweetness and Power – Public policies and the 'biofuels frenzy'. *EuroChoices*, 11(3), 23–28.

Adger, W.N. and Jordan, A. (eds.) (2009) *Governing Sustainability*. Cambridge: Cambridge University Press.

Afionis, S. and Stringer, L.C. (2012) European Union Leadership in Biofuels Regulation: Europe as a normative power? *Journal of Cleaner Production*, 32, 114–123.

Algieri, B. (2014) The Influence of Biofuels, Economic and Financial Factors on Daily Returns of Commodity Futures Prices. *Energy Policy*, 69, 227–247.

Andrade de Sá, S., Palmer, C. and dui Falco, S. (2013) Dynamics of Indirect Land-Use Change: Empirical evidence from Brazil. *Journal of Environmental Economics and Management*, 65, 377–393.

Auld, D. (2013) *Accountability Denied: The Global Biofuel Blunder*. Victoria, BC: FriesenPress.

Babcock, B. (2011) *The Impact of US Biofuel Policies on Agricultural Price Levels and Volatility*. International Centre for Trade and Sustainable Development (ICTSD) Programme on Agricultural Trade and Sustainable Development Issue Paper No. 17. Available at: http://ictsd.org/downloads/2011/12/the-impact-of-us-biofuel-policies-on-agricultural-price-levels-and-volatility.pdf

Baffes, J. and Dennis, A. (2013) *Long-Term Drivers of Food Prices*. Policy Research Working Paper 6455. The World Bank Development Prospects Group & Poverty Reduction and Economic Management Network Trade Department. Available at: http://elibrary.worldbank.org/doi/pdf/10.1596/1813-9450-6455

Bakvis, H. (2000) Rebuilding Policy Capacity in the Era of the Fiscal Dividend: A report from Canada. *Governance*, 13, 71–103.

Bakvis, H. and Aucoin, P. (2005) Public Service Reform and Policy Capacity: Recruiting and Retaining the Best and the Brightest? In Painter, M. and Pierre, J. (eds.) *Challenges To State Policy Capacity*. Basingstoke: Palgrave, 185–204.

Banse, M., van Meijl, H. and Woltjer, G. (2008) Consequences of EU Biofuel Policies on Agricultural Production and Land Use. *Choices*, 23(3), 22–27.

Barnes, P.M. (2011) The Role of the Commission of the European Union: Creating External Coherence from Internal Diversity. In Wurzel, R. and Connelly, J. (eds.) *The European Union as a Leader in International Climate Change Politics*. UACES Contemporary European Studies Series. London: Routledge for the University Association for Contemporary European Studies, 42–57.

Barrett, C.B. and M.F. Bellemare (2011) *The G-20's Error: Food Price Volatility is Not the Problem*. Available at: http://dyson.cornell.edu/faculty_sites/cbb2/Papers/BarrettBellemareFoodPricesJune2011.pdf

Barrett, S. (2004) Implementation Studies: Time for a revival? Personal reflections on 20 years of implementation studies. *Public Administration*, 82, 249–262.

Bastianin, A., Galeotti, M. and Manera, M. (2013) *Biofuels and Food Prices: Searching for the Causal Link*. Fondazione Eni Enrico Mattei Nota di Lavoro 22.2013. Available at: http://www.feem.it/getpage.aspx?id=5393&sez=Publications&padre=73

Bastos Lima, M.G. and Gupta, J. (2013) The Policy Context of Biofuels: A case of non-governance at the global level? *Global Environmental Politics*, 13(2), 46–64.

Berazneva, J. and Lee, D.R. (2013) Explaining the African Food Riots of 2007–2008: An empirical analysis. *Food Policy*, 39, 28–39.

Bevill, K. (2008) Building the 'Minnesota Model'. *Ethanol Producer Magazine*, April, 114–120.

Bocquillon, P. and Dobbels, M. (2014) An Elephant on the 12th Floor of the Berlaymont? European council and commission relations in legislative agenda setting. *Journal of European Public Policy*, 21(1), 20–38.

Bomb, C., McCormick, K., Deurwaarder, E. and Kåberger, T. (2007) Biofuels for Transport in Europe: Lessons from Germany and the UK. *Energy Policy*, 35, 2256–2267.

Borrell, B. and Duncan, R.C. (1993) A Survey of World Sugar Policies. In Marks, S.V. and Maskus, K.E. (eds.) *The Economics and Politics of World Sugar Policies*. Ann Arbor: The University of Michigan Press, 15–48.

Bourblanc, M. (2011) The European Union and Global Environmental Governance. In Wunderlich, J.-U. and Bailey, D.J. (eds.) *The European Union and Global Governance: A Handbook*. Routledge International Handbooks. London: Routledge, 131–139.

Bovens, M., ''t Hart', P. and Peters, B.G. (2001) *Success and Failure in Public Governance: A Comparative Analysis*. Cheltenham: Edward Elgar.

BP (2014) *BP Energy Outlook 2035*. Available at: www.bp.com/energyoutlook, accessed 1 February 2014.

Broch, A., Hoekman, S.K. and Unnasch, S. (2013) A Review of Variability in Indirect Land Use Change Assessment and Modeling in Biofuel Policy. *Environmental Science and Policy*, 29, 147–157.

Brown, L.R. (2012) *Full Planet, Empty Plates: The New Geopolitics of Food Security*. New York: W.W. Norton & Company.

Brown, R.C. and Brown, T.R. (2012) *Why Are We Producing Biofuels?: Shifting to the Ultimate Source of Energy*. Ames, IA: Brownia LLC.

Brunelle, T. and Dumas, P. (2012) *Can Numerical Models Estimate Indirect Land-Use Change?* Fondazione Eni Enrico Mattei, Nota di Lavoro 65.2012. Available at: http://www.feem.it/userfiles/attach/2012913123274NDL2012-065.pdf

Busse, S., Brümmer, B. and Ihle, R. (2010) *The Pattern of Integration Between Fossil Fuel and Vegetable Oil Markets: The case of biodiesel in Germany.* Paper presented to the Agricultural and Applied Economics Association meeting, Denver, July 2010. Available at: http://ageconsearch.umn.edu/bitstream/61010/2/Busse_AAEA.pdf

Cacciatore, M.A., Binder, A.R., Scheufele, D.A. and Shaw, B.R. (2012a) Public Attitudes Toward Biofuels: Effects of knowledge, political partisanship, and media use. *Politics and the Life Sciences*, 31(1–2), 36–51.

Cacciatore, M.A., Scheufele, D.A. and Shaw, B.R. (2012b) Labeling Renewable Energies: How the language surrounding biofuels can influence its public acceptance. *Energy Policy*, 51, 673–682.

Carvalho, L.C.C. (1999) Prospects for Ethanol and Sugar in Brazil. *International Sugar Journal (Cane Sugar Edition)*, 101(1211), 574–577.

CBES (2009) *Land-Use Change and Bioenergy: Report from the 2009 Workshop.* ORNL/CBES-001, U.S. Department of Energy, Office of Energy Efficiency and Renewable Energy and Oak Ridge National Laboratory, Center for BioEnergy Sustainability. Available at: http://web.ornl.gov/sci/ees/cbes/Publications/CBES1_LUCB_rpt_FINAL4.pdf

César, A.S. and Batalha, M.O. (2013) Brazilian Biodiesel: The case of the palm's social projects. *Energy Policy*, 56, 165–174.

CGEE (2012) *Sustainability of Sugarcane Bioenergy*, updated edition. Centro de Gestão e Estudos Estratégicos, Brasilia. Available at: http://www.cgee.org.br/publicacoes/sustainability.php

Chalabi, F.J. (2010) *Oil Policies, Oil Myths: Analysis and Memoir of an OPEC 'Insider'.* London: I. B. Tauris.

Charles, M.B., Ryan, R., Ryan, N. and Oloruntoba, R. (2007) Public Policy and Biofuels: The way forward? *Energy Policy*, 35, 5737–5746.

Charnovitz, S., Earley, J. and Howse, R. (2008) *An Examination of Social Standards in Biofuels Sustainability Criteria.* International Food and Agricultural Trade Policy Council (IPC) Standards Series. Available at: http://www.agritrade.org/documents/SocialStnds_Biofuels_FINAL.pdf

Cheyne, I. (2009) Proportionality, Proximity and Environmental Labelling in WTO Law. *Journal of International Economic Law*, 12, 927–952.

Chin, H.-C., Choong, W.-W., Wan Alwi, S.R. and Mohammed, A.H. (2014) Issues of Social Acceptance on Biofuel Development. *Journal of Cleaner Production.* Article in Press, http://dx.doi.org/10.1016/j.jclepro.2013.12.060.

Chite, R.M. (2012) *Previewing the Next Farm Bill.* Congressional Research Service Paper R42357. Available at: http://www.fas.org/sgp/crs/misc/R42357.pdf.

Ciaian, P. and Kancs, d'A. (2011) Interdependencies in the Energy-Bioenergy-Food Price Systems: A cointegration analysis. *Resource and Energy Economics*, 33, 326–348. An earlier version of this research is available freely at: http://www.eeri.eu/documents/wp/EERI_RP_2009_06.pdf

Clancy, J. (21013) *Biofuels and Rural Poverty.* Abingdon: Routledge.

Colares, J.F. (2008) A Brief History of Brazilian Biofuels Legislation. *Syracuse Journal of International Law and Commerce*, 35(2), 293–308.

Commission of the European Communities (2001a) *Communication from the Commission: A sustainable Europe for a better world: A European union strategy for sustainable development*. COM(2001)264 final, Brussels, 15.5.2001.

Commission of the European Communities (2001b) *Communication from the Commission ... on Alternative Fuels for Road Transportation and on a Set of Measures to Promote the Use of Biofuels; Proposal for a directive ... on the promotion of the use of biofuels in transport; Proposal for a council directive ... with regard to the possibility of applying a reduced rate of excise duty on certain mineral oils containing biofuels and on biofuels*. COM(2001)547 final, Brussels, 7.11.2001.

Commission of the European Communities (2002) *Amended Proposal for a Directive ... on the Promotion of the use of biofuels for transport*. COM(2002)508 final, Brussels, 12.9.2002.

Commission of the European Communities (2004) *Communication from the Commission ... The share of renewable energy in the EU*. COM(2004)366 final, Brussels, 26.5.2004.

Commission of the European Communities (2005a) *Green Paper on Energy Efficiency or Doing More with Less*. COM(2005)265 final, Brussels, 22.6.2005.

Commission of the European Communities (2005b) *Communication from the Commission: The support of electricity from renewable energy sources*. COM(2005)627 final, Brussels, 7.12.2005.

Commission of the European Communities (2005c) *Communication from the Commission: Biomass action plan*. COM(2005)628 final, Brussels, 7.12.2005.

Commission of the European Communities (2006a) *Communication from the Commission: An EU strategy for biofuels*. COM(2006)34 final, Brussels, 8.2.2006.

Commission of the European Communities (2006b) *Green Paper: A European strategy for sustainable, competitive and secure energy*. COM(2006)105 final, Brussels, 8.3.2006.

Commission of the European Communities (2007a) *Communication from the Commission ... Biofuels Progress Report*. COM(2006)845 final, Brussels, 10.1.2007.

Commission of the European Communities (2007b) *Communication from the Commission ... Renewable Energy Road Map*. COM(2006)848 final, Brussels, 10.1.2007.

Commission of the European Communities (2007c) *Commission Staff Working Document Accompanying the Renewable Energy Road Map: Impact assessment*. SEC(2006)1719, Brussels, 10.1.2007.

Commission of the European Communities (2007d) *Commission Staff Working Document Accompanying the Biofuels Progress Report: Review of economic and environmental data for the biofuels progress report*. SEC(2006)1721, Brussels, 10.1.2007.

Condon, B.J. (2009) Climate Change and Unresolved Issues in WTO Law. *Journal of International Economic Law*, 12, 895–926.

Congressional Budget Office (2010) *Using Biofuel Tax Credits to Achieve Energy and Environmental Policy Goals*. The Congress of the United States: Congressional Budget Office.

Copenhagen Economics (2011) *The Missing Indirect Land-Use Change Factors: How to make decisions when science is incomplete?* Report for the European Forum for Sustainable Development. Available at (last accessed 12 February

2014): http://www.copenhageneconomics.com/Files/Filer/Publikationer/142201%20Copenhagen%20Economics%20-%20Indirect%20land%20use%20change.pdf

Daugbjerg, C. and Swinbank, A. (forthcoming) Globalization and New Policy Concerns: The WTO and the EU's sustainability criteria for biofuels. *Journal of European Public Policy.* DOI:10.1080/13501763.2014.927520

de Beer, J. and Smyth, S.J. (2012) International Trade in Biofuels: Legal and regulatory issues. *The Estey Centre Journal of International Law and Trade Policy*, 13(1), 131–149. Available at: http://ageconsearch.umn.edu/handle/125041

DeCicco, J.M. (2013) Biofuel's Carbon Balance: Doubts, certainties and implications. *Climatic Change*, 121, 801–814. Available on open access at: http://link.springer.com/article/10.1007%2Fs10584-013-0927-9

de Gorter, H. and Just, D.R. (2009) The Economics of a Blend Mandate for Biofuels. *American Journal of Agricultural Economics*, 91, 738–750.

Delshad, A. and Raymond, L. (2013) Media Framing and Public Attitudes Toward Biofuels. *Review of Policy Research*, 30(2), 190–210.

Delshad, A.B., Raymond, L., Sawicki, V. and Wegener, D.T. (2010) Public Attitudes Toward Political and Technological Options for Biofuels. *Energy Policy*, 38, 3414–3425.

Delzeit, R., Klepper, G. and Lange, M. (undated) Review of IFPRI Study 'Assessing the Land Use Change Consequences of European Biofuel Policies and Its Uncertainties'. Study on behalf of the European biodiesel board (Accessed 3 January 2013): http://www.ebb-eu.org/EBBpressreleases/Review_iLUC_IfW_final.pdf

Deshpande, R.S. (2006) Biofuels and WTO: An emerging context. *Asian Biotechnology and Development Review*, 8(2), 77–96. Available at: http://ris.org.in/images/RIS_images/pdf/article5_v8n2.pdf

Desplechin, E. (2010) The Brazilian Sugarcane Ethanol Industry's Approach to Certification. *Sugar Industry*, 135, 222–225.

Devadoss, S. and Kuffel, M. (2010) Is the U.S. Import Tariff on Brazilian Ethanol Justifiable? *Journal of Agricultural and Resource Economics*, 35(3), 476–488.

de Vera, E.R. (2008) The WTO and Biofuels: The possibility of unilateral sustainability requirements. *Chicago Journal of International Law*, 8, 661–679.

Di Lucia, L., Ahlgren, S. and Ericsson, K. (2012) The Dilemma of Indirect Land-Use Changes in EU Biofuel Policy – An empirical study of policy-making in the context of scientific uncertainty. *Environmental Science and Policy*, 16, 9–19.

Di Lucia, L. and Kronsell, A. (2010) The Willing, the Unwilling and the Unable – Explaining implementation of the EU biofuels directive. *Journal of European Public Policy*, 17(4), 545–563.

Dollive, K. (2008) *The Impact of Export Restraints on Rising Grain Prices.* Office of economics working paper 2008–09-A. US international trade commission. Available at: http://www.usitc.gov/publications/332/working_papers/EC200809A.pdf

Drèze, J. and Sen, A. (1989) *Hunger and Public Action. WIDER Studies in Development Economics.* Oxford: Oxford University Press/Clarendon Press.

Du, X. and McPhail, L.L. (2012) Inside the Black Box: The price linkage and transmission between energy and agricultural markets. *The Energy Journal*, 33(2), 171–194. An earlier version of this research is available freely at: http://ageconsearch.umn.edu/bitstream/103268/2/Black%20Box%20edit.pdf

Du, X., Yu, C.L. and Hayes, D.J. (2011) Speculation and Volatility Spillover in the Crude Oil and Agricultural Commodity Markets: A Bayesian analysis. *Energy Economics*, 33, 497–503. An earlier version of this research is available freely at: http://www.card.iastate.edu/publications/DBS/PDFFiles/ 09wp491.pdf?origin=publication_detail

Dumortier, J., Hayes, D.J., Carriquiry, M., Dong, F., Du, X., Elobeid, A., Fabiosa, J.F. and Tokgoz, S. (2011) Sensitivity of Carbon Emission Estimates from Indirect Land-Use Change. *Applied Economic Perspectives and Policy*, 33(3), 428–448.

Echols, M.A. (2009) *Biofuels Certification and the Law of the World Trade Organization*. ICTSD Global Platform on Climate Change, Trade Policies and Sustainable Energy Issue Paper No. 17. Geneva: International Centre for Trade and Sustainable Development. Available at: http://ictsd.org/i/publications/53937/ ?view=document

Edwards, R., Mulligan, D. and Marelli, L. (2010) *Indirect Land Use Change From Increased Biofuels Demand: Comparison of models and results from marginal biofuels production from different feedstocks*. Joint Research Centre Scientific and Technical Report JRC 59771, EUR 24485 EN. Available at: http://ec.europa.eu/energy/renewables/studies/doc/land_use_change/ study_4_iluc_modelling_comparison.pdf

Edwards, R., Szekeres, S., Neuwahl, F. and Mahieu, V. (2008) *Biofuels in the European Context: Facts, uncertainties and recommendations*. European Commission Joint Research Centre Scientific and Technical Reports JRC 43285. Luxembourg: Office for Official Publications of the European Communities. Available at: <http://www.energy.eu/publications/LBNA23260ENC_002.pdf>

Elmarzougui, E. and Larue, B. (2013) On the Evolving Relationship Between Corn and Oil Prices. *Agribusiness: An International Journal*, 29(3), 344–360. An earlier version of this research is available freely at: http://www.create.ulaval.ca/ uploads/tx_centrerecherche/CREATE2011-3.pdf

Energy Future Coalition (2007) *Biofuels For Our Future: A primer*. Washington, DC: Energy Future Coalition. Available at: <http://www.globalproblems-globalsolutions-files.org/unf_website/PDF/biofuels_faq.pdf>.

Environmental Working Group (2011) *Corn Ethanol Subsidy: A losing proposition*. Available at: http://www.ewg.org/agmag/2011/03/corn-ethanol-subsidy-losing-proposition, accessed 1 December 2013.

Erickson, B., Carr, M. and Winters, P. (2014) Estimating Greenhouse Gas Emissions from Proposed Changes to the Renewable Fuel Standard Through 2022. *Industrial Biotechnology*, 10(2), 57–63. Available at: http://online.liebertpub. com/doi/full/10.1089/ind.2014.1508

Erixon, F. (2009) Green Protectionism in the European Union: How Europe's biofuels policy and the renewable energy directive violate WTO commitments. ECIPE occasional paper 1/2009. European centre for international political economy. Available at: http://www.ecipe.org/green-protectionism-in-the-european-union-how-europe2019s-biofuels-policy-and-the-renewable-energy-directive-violate-wto-commitments/PDF

EU (2003) Directive 2003/30/EC of the European Parliament and of the Council of 8 May 2003 on the Promotion of the Use of Biofuels or Other Renewable Fuels for Transport (Biofuels Directive). *Official Journal of the European Union*, L123, 17.5.2003, 42–46.

EU (2009a) Directive 2009/28/EC of the European Parliament and of the Council of 23 April 2009 on the Promotion of The Use of Energy from Renewable Sources and Amending and Subsequently Repealing Directives 2001/77/EC and 2003/30/EC (Renewable Energy Directive). *Official Journal of the European Union*, L140, 5.6.2009, 16–62.

EU (2009b) Directive 2009/30/EC of the European Parliament and of the Council of 23 April 2009 Amending Directive 98/70/EC as Regards the Specification of Petrol, Diesel and Gas-Oil and Introducing a Mechanism to Monitor and Reduce Greenhouse Gas Emissions....*Official Journal of the European Union* L140, 5.6.2009, 88–113.

European Commission (1997) *Energy for the Future: Renewable sources of energy. White paper for a community strategy and action plan.* Communication from the Commission, COM(97)599 final.

European Commission (2010) *Report from the Commission: On indirect land-use change related to biofuels and bioliquids.* COM(2010)811 final, Brussels, 22.12.2010.

European Commission (2012) *Proposal for a Directive...[amending the FQD and the RED].* COM(2012)595 final, Brussels, 17.10.2012.

European Commission (2014) *Guidelines on State Aid for Environmental Protection and Energy 2014–2020.* At the Time of Writing the Final Version of the Document Had Not Been Prepared, Nor a Document Number Designated. The Preliminary Version is Available at: http://ec.europa.eu/competition/sectors/energy/eeag_en.pdf (accessed 19 April 2014).

FAO (2008) *The State of Food and Agriculture 2008.* Rome: Food and Agriculture Organization of the United Nations.

FAO (2009) *The State of Agricultural Commodity Markets: High food prices and the food crisis – Experiences and lessons learned.* Rome: Food and Agriculture Organization of the United Nations. Available at: ftp://ftp.fao.org/docrep/fao/012/i0854e/i0854e.pdf

FAO (2012) *Biofuels Co-Products as Livestock Feed: Opportunities and challenges.* Rome: Food and Agriculture Organization of the United Nations. Available at: http://www.fao.org/docrep/016/i3009e/i3009e.pdf

Fargione, J., Hill, J., Tilman, D., Polasky, S. and Hawthorne, P. (2008) Land Clearing and the Biofuel Carbon Debt. *Science*, 319, 1235–1238.

Farinelli, B., Carter, C.A., Lin, C.-Y.C. and Sumner, D.A. (2009) Import Demand for Brazilian Ethanol: A cross-country analysis. *Journal of Cleaner Production*, 17, S9–S17.

Financial Times (2013a) US Ethanol Biofuels Mix Hits the Oil 'Blendwall'. 14 August.

Financial Times (2013b) Biofuels: US steers bumpy course to green fuel. 3 June.

Frondel, M. and Peters, J. (2007) Biodiesel: A new oildorado? *Energy Policy*, 35, 1675–1684.

Gamborg, C., Anker, H.T. and Sandøe, P. (2014) Ethical and Legal Challenges in Bioenergy Governance: Coping with value disagreement and regulatory complexity. *Energy Policy*, 69, 326–333.

Garcez, C.A.G and Vianna, J.N.S. (2009) Brazilian Biodiesel Policy: Social and environmental considerations of sustainability. *Energy*, 34(5), 645–654.

Gardebroek, C. and Hernandez, M.A. (2013) Do Energy Prices Stimulate Food Price Volatility? Examining volatility transmission between us oil, ethanol and corn markets. *Energy Economics*, 40, 119–129. An earlier version of this research Is Available freely at: http://ageconsearch.umn.edu/bitstream/124583/2/draft_energy_corn_CG_MH_june_1_2012_AAEA.pdf

Gee, S. and McMeekin, A. (2011) Eco-Innovation Systems and Problem Sequences: The contrasting cases of US and Brazilian biofuels. *Industry and Innovation*, 18(3), 301–315.

Gilbertson, T., Holland, N., Semino, S. and Smith, K. (2008) *Paving the Way for Agrofuels: EU policy, sustainability criteria and climate calculations*, 2nd revised printing. Discussion Paper 2007. Amsterdam: Transnational Institute; Corporate Europe Observatory; Grupo do Reflexión Rural.

Gillon, S. (2014) Science in Carbon Economies: Debating what counts in US biofuel governance. *Environment and Planning A*, 46(2), 318–336.

Goldemberg, J. (2006) The Ethanol Program in Brazil. *Environmental Research Letters*, 1, 5pp (article pages not numbered sequentially in this journal).

Goodwin, B.K., Marra, M., Piggott, N. and Müeller, S. (2012) *Is Yield Endogenous to Price? An empirical evaluation of inter- and intra-seasonal corn yield response*. Paper delivered to the Agricultural and Applied Economics Association 2012 Annual Meeting, August 12–14, 2012, Seattle, Washington. Available at: http://ageconsearch.umn.edu/handle/124884

Grossman, P.Z. (2012) The Logic of Deflective Action: US energy shocks and the US policy process. *Journal of Public Policy*, 32(1), 33–51.

Guan, E.J. (2010) Understanding Brazil's Oil Industry: Policy dynamics and self-sufficiency. *Journal of Emerging Knowledge on Emerging Markets*, 2, 76–98.

Harmer, T. (2009) *Biofuels Subsidies and the Law of the WTO*. ICTSD Global Platform on Climate Change, Trade Policies and Sustainable Energy Issue Paper No. 20. Geneva: International centre for trade and sustainable development. Available at: http://ictsd.org/downloads/2012/03/biofuels-subsidies-and-the-law-of-the-wto.pdf

Harvey, M. and McMeekin, A. (2010) *Political Shaping of Transitions to Biofuels in Europe, Brazil and the USA*. Centre for Research in Economic Sociology and Innovation (CRESI) Working Paper 2010–02, University of Essex. Available at: http://repository.essex.ac.uk/2296/1/CWP-2010-02-Political-Shaping-Final.pdf

Havlík, P. [and 14 others] (2011) Global Land-Use Implications of First and Second Generation Biofuel Targets. *Energy Policy*, 39, 5690–5702.

Helfand, S. M. (1999) The Political Economy of Agricultural Policy in Brazil: Decision making and influence from 1964 to 1992. *Latin American Research Review*, 34(2), 3–41.

Hertel, T.W., Tyner, W.E. and Birur, D.K. (2010) The Global Impacts of Biofuel Mandates. *The Energy Journal*, 31(1), 75–100.

Hira, A. and Guilherme de Oliveira, L. (2009) No Substitute for Oil? How Brazil developed its ethanol industry. *Energy Policy*, 37, 2450–2456.

Horta Nogueira, L.A. and Silva Capaz, R. (2013) Biofuels in Brazil: Achievements and perspectives on food security. *Global Food Security*, 2, 117–125.

Howes, T. (2010) The EU's New Renewable Energy Directive (2009/28/EC). In Oberthür, S. and Pallemaerts, M. (eds.) *The New Climate Policies of the European Union*. Brussels: VUB Press (Brussels University Press), 117–150.

Howlett, M. (2009) Governance Modes, Policy Regimes and Operational Plans: A multi-level nested model of policy instrument choice and policy design. *Policy Sciences*, 42(1), 73–89.

Howlett, M. and Lindquist, E. (2004) Policy Analysis and Governance: Analytical and policy styles in Canada. *Journal of Comparative Policy Analysis*, 6, 225–249.

Howlett, M. and Rayner, J. (2006a) Understanding the Historical Turn in the Policy Sciences: A critique of stochastic, narrative, path dependency and process-sequencing models of policy-making over time. *Policy Sciences*, 39, 1–18.

Howlett, M. and Rayner, J. (2006b) Globalization and Governance Capacity: Explaining divergence in national forest programs as instances of 'next-generation' regulation in Canada and Europe. *Governance*, 19(2), 251–275.

Howlett, M. and Rayner, J. (2007) Design Principles for Policy Mixes: Cohesion and coherence in 'New Governance Arrangements'. *Policy and Society*, 26(4), 1–18.

Howse, R., van Bork, P. and Hebebrand, C. (2006) *WTO Disciplines and Biofuels: Opportunities and constraints in the creation of a global marketplace*. International Food and Agricultural Trade Policy Council (IPC) Discussion Paper. Available at: http://www.agritrade.org/Publications/DiscussionPapers/WTO_Disciplines_Biofuels.pdf

Huchet-Bourdon, M. (2011) *Agricultural Commodity Price Volatility: An overview*. OECD Food, Agriculture and Fisheries Working Papers, No. 52. Paris: Organisation for Economic Cooperation and Development. Available at: http://www.oecd-ilibrary.org/agriculture-and-food/agricultural-commodity-price-volatility_5kg0t00nrthc-en

Hutchinson, H.W. (1961) The Transformation of Brazilian Plantation Society. *Journal of Inter-American Studies*, 3(2), 201–212.

IEA Bioenergy (2011) *Bioenergy, Land Use Change and Climate Change Mitigation: Background technical report*. IEA Bioenergy:ExCo:2011:04. Available at: http://www.ieabioenergy.com/wp-content/uploads/2013/10/Bioenergy-Land-Use-Change-and-Climate-Change-Mitigation-Background-Technical-Report.pdf

Informa Economics (2011) *Analysis of Corn, Commodity, and Consumer Food Prices*. Informa Economics, prepared for the Renewable Fuels Foundation. Available at: http://ethanolrfa.org/page/-/Exec%20Summ%20and%20Concl.%20Consumer%20Food%20Price%20Analysis%20Report%20%28July%202011%29.pdf?nocdn=1&utm_medium=email&utm_campaign=Informa%20Study&utm_content=Informa%20Study+CID_b610d05d34dd6aecfe91adcdcaa3a640&utm_source=Email+marketing+software&utm_term=comprehensive+study

Jacobs, A.M. (2011) *Governing for the Long Term*. New York: Cambridge University Press.

Jacoby, W. and Meunier, S. (2010) Europe and the Management of Globalization. *Journal of European Public Policy*, 17(3), 299–317.

Jensen, M. and Andersen, A.H. (2013) Biofuels: A contested response to climate change. *Sustainability: Science, Practice, and Policy*, 9(1), 42–56.

Johnson, F.I. (1983) Sugar in Brazil: Policy and production. *The Journal of Developing Areas*, 17, 243–256.

Johnston, M., Licker, R., Foley, J., Holloway, T., Mueller, N.D., Carford, C. and Kucharik, C. (2011) Closing the Gap: Global potential for increasing biofuel production through agricultural intensification. *Environmental Research Letters*, 6(3). Available at: http://iopscience.iop.org/1748-9326/6/3/034028/pdf/1748-9326_6_3_034028.pdf

Josling, T., Blandford, D. and Earley, J. (2010) *Biofuel and Biomass Subsidies in the U.S., EU and Brazil: Towards a transparent system of notification*. International Food and Agricultural Trade Policy Council (IPC) position paper. Available at: http://www.agritrade.org/documents/Biofuels_Subs_Web_Final.pdf

Kavalov, B. (2004) *Biofuels Potentials in the EU*. Report EUR 21012 EN, European Commission Joint Research Centre.

Kay, A. and Ackrill, R. (2009) Institutional Change in the International Governance of Agriculture: A revised account. *Governance*, 22(3), 483–506.

Kay, A. and Ackrill, R. (2012) Governing the Transition to a Biofuels Economy in the US and EU: Accommodating value conflicts, implementing uncertainty. *Policy and Society*, 31, 295–306.

Kelemen, R.D. (2010) Globalizing European Union Environmental Policy. *Journal of European Public Policy*, 17(3), 335–349.

Kim, S. and Dale, B.E. (2011) Indirect Land Use Change for Biofuels: Testing predictions and improving analytical methodologies. *Biomass and Bioenergy*, 35, 3235–3240.

Kobrin, S.J. (2009) Sovereignty @ Bay. In A.M. Rugman (ed.) *The Oxford Handbook of International Business*, 2nd ed. Oxford: Oxford University Press, 183–204.

Koppenjan, J. and Klijn, E.-H. (2004) *Managing Uncertainties in Networks: A network approach to problem solving and decision making*. London: Routledge.

Kretschmer, B., Bowyer, C. and Buckwell, A. (2012) *EU Biofuels Use and Agricultural Commodity Prices: A review of the evidence base*. Report prepared for Action Aid. Institute for European Environmental Policy. Available at: http://www.globalbioenergy.org/uploads/media/1206_IEEP_-_EU_biofuel_use_and_agricultural_commodity_prices_a_review_of_the_evidence_base.pdf

Kretschmer, B., Daiju, N. and Peterson, S. (2009) The Economic Effects of the EU Biofuel Target. *Energy Economics*, 31 (Supplement 2), S285–S294.

Kutas, G., Lindberg, C. and Steenblik, R. (2007) *Biofuels: At What Cost? Government support for ethanol and biodiesel in the European Union*. Report for the Global Subsidies Initiative, International Institute for Sustainable Development. Geneva: IISD.

Labruto, N. (2014) Experimental Biofuel Governance: Historicizing social certification in Brazilian ethanol production. *Geoforum*. Article in Press. Available at: http://www.sciencedirect.com/science/article/pii/S0016718514000530

Lamers, P., Hamelinck, C., Junginger, M. and Faaij, A. (2011) International Bioenergy Trade – A review of past developments in the liquid biofuel market. *Renewable and Sustainable Energy Reviews*, 15, 3655–3676.

Langeveld, H., Sanders, J. and Meeusen, M. (eds.) (2010) *The Biobased Economy: Biofuels, materials and chemicals in the post-oil era*. London: Earthscan.

Langeveld, J.W.A., Dixon, J., van Keulen, H. and Quist-Wessel, P.M.F. (2013) Analyzing the Effect of Biofuel Expansion on Land Use in Major Producing Countries: Evidence of increased multiple cropping. *Biofuels, Bioproducts and Biorefining*, 8, 49–58. NB: A longer, publicly-available, report is available at: http://www.biomassresearch.eu/Biomass%20Research%201301_Analysing%20the%20effect%20of%20biofuel%20expansion%20on%20land%20use.pdf

Lankoski, J. and Ollikainen, M. (2011) Biofuel Policies and the Environment: Do climate benefits warrant increased production from biofuel feedstocks? *Ecological Economics*, 70, 676–687.

Le Roy, D.G., Elobeid, A.E. and Klein, K.K. (2011) The Impact of Trade Barriers on Mandated Biofuel Consumption in Canada. *Canadian Journal of Agricultural Economics*, 59(4), 457–74. An earlier, 2009, version is also available freely at: http://ageconsearch.umn.edu/bitstream/54973/2/Leroy-TPB%202009-2.pdf

Lehtonen, M. (2011) Social Sustainability of the Brazilian Bioethanol [sic]: Power relations in a centre-periphery perspective. *Biomass and Bioenergy*, 35, 2425–34.

Lendle, A. and Schaus, M. (2010) *Sustainability Criteria in the EU Renewable Energy Directive: Consistent with WTO rules?*. International Centre for Trade and Sustainable Development (ICTSD) Information Note No. 2. Available at: http://ictsd.org/downloads/2011/12/sustainability-criteria-in-the-eu-renewable-energy-directive-consistent-with-wto-rules.pdf

Liapis, P. (2012) *Structural Change in Commodity Markets: Have agricultural markets become thinner?* OECD Food, Agriculture and Fisheries Working Papers, No. 54. Paris: Organisation for Economic Cooperation and Development. Available at: http://www.oecd-ilibrary.org/agriculture-and-food/structural-change-in-commodity-markets_5k9fp3zdc1d0-en

Liew, W.H., Hassim, M.H. and Ng, D.K.S. (2014) Review of Evolution, Technology and Sustainability Assessments of Biofuel Production. *Journal of Cleaner Production*. Article in Press, http://dx.doi.org/10.1016/j.jclepro.2014.01.006.

Loppacher, L.J. and Kerr, W.A. (2005) Can Biofuels become a Global Industry? Government policies and trade constraints. *Energy Politics*, V (Spring), 7–27.

Lydgate, E.B. (2012a) Sustainable Development in the WTO: From mutual supportiveness to balancing. *World Trade Review*, 11(4), 621–39.

Lydgate, E.B. (2012b) Biofuels, Sustainability and Trade-Related Regulatory Chill. *Journal of International Economic Law*, 15(1), 157–80.

Mahoney, J. and Thelen, K. (2010) A Theory of Gradual Institutional Change. In Mahoney, J. and Thelen, K. (eds.) *Explaining Institutional Change: Ambiguity, agency, and power*. New York: Cambridge University Press, 1–37.

March, J. (1978) Bounded Rationality, Ambiguity, and the Engineering of Choice. *The Bell Journal of Economics*, 9, 587–608.

Maréchal, K. (2009) An Evolutionary Perspective on the Economics of Energy Consumption: The crucial role of habits. *Journal of Economic Issues*, XLIII(1), 69–88.

Martines-Filho, J., Burnquist, H.L. and Vian, C.E.F. (2006) Bioenergy and the Rise of Sugarcane-Based Ethanol in Brazil. *Choices*, 21(2), 91–96.

Martins Borges, J.M. (1986) Proálcool: A success story. *World Commodity Journal*, IX(415), 28–29.

Matthews, F. (2012) Governance and State Capacity. In D. Levi-Faur (ed.) *The Oxford Handbook of Governance*. Oxford: Oxford University Press, 281–93.

May, P.J. (1992) Policy Learning and Failure. *Journal of Public Policy*, 12(4), 331–54.

May, P.J., Jones, B.D., Beem, B.E., Neff-Sharum, E.A. and Poague, M.K. (2005) Policy Coherence and Component-Driven Policymaking: Arctic policy in Canada and the United States. *Policy Studies Journal*, 33(1), 37–63.

McCarthy, J.E. and Tiemann, M. (2006) *MTBE in Gasoline: Clean air and drinking water issues*. CRS Report for Congress RL32787. Available at: http://www.cnie.org/nle/crsreports/06may/RL32787.pdf

McPhail, L.L., Du, X. and Muhammad, A. (2012) Disentangling Corn Price Volatility: The role of global demand, speculation, and energy. *Journal of Agricultural and Applied Economics*, 44(3), 401–10.

Melillo, J.M. [and 9 others] (2009) Indirect Emissions from Biofuels: How important? *Science*, 326, 1397–99.

Meyer, S., Schmidhuber, J. and Barreiro-Hurlé, J. (2013) *Global Biofuel Trade: How uncoordinated biofuel policy fuels resource use and GHG emissions*. International Centre for Trade and Sustainable Development (ICTSD) Programme on Agricultural Trade and Sustainable Development Issue Paper No. 48. Available at: http://ictsd.org/downloads/2013/05/global-biofuel-trade-how-uncoordinated-biofuel-policy-fuels-resource-use-and-ghg-emissions.pdf

Milne, J.L. and Field, C.B. (2012) Assessment Report from the GCEP Workshop on Energy Supply with Negative Carbon Emissions. Global Climate and Energy Project. Available at: https://gcep.stanford.edu/events/workshops_negemissions2012.html

Mitchell, A. and Tran, C. (2010) The Consistency of the European Union Renewable Energy Directive with World Trade Organization Agreements: The case of biofuels. *Renewable Energy Law and Policy Review*, 1/2010, 33–44. An earlier version of this is available freely at: http://scholarship.law.georgetown.edu/fwps_papers/119/

Mitchell, D. (2008) *A Note of Rising Food Prices*. Policy Research Working Paper 4682. The World Bank Development Prospects Group. Available at: http://www-wds.worldbank.org/servlet/WDSContentServer/WDSP/IB/2008/07/28/000020439_20080728103002/Rendered/PDF/WP4682.pdf

Mitchell, D. (2011) *Biofuels in Africa: Opportunities, prospects, and challenges*. Washington, DC: The World Bank.

Mohr, A. and Raman, S. (2013) Lessons from First Generation Biofuels and Implications for the Sustainability Appraisal of Second Generation Biofuels. *Energy Policy*, 63, 114–22.

Moïsé, E. and Steenblik, R. (2011) *Trade-Related Measures Based on Processes and Production Methods in the Context of Climate-Change Mitigation*. OECD Trade and Environment Working Papers 2011/04. Available at: http://www.oecd-ilibrary.org/trade/trade-related-measures-based-on-processes-and-production-methods-in-the-context-of-climate-change-mitigation_5kg6xssz26jg-en

Mondou, M., Skogstad, G. and Houle, D. (2014) Policy Image Resilience, Multidimensionality, and Policy Image Management: A study of US biofuel policy. *Journal of Public Policy*, 34(1), 155–80.

Morceli, P. (2007) Brazilian Alcohol: prospects. *Revista de Politica Agricola*, Special Edition (English version of volume 3, 2006), 19–27. http://www.agricultura.gov.br/pls/portal/docs/PAGE/MAPA/MENU_LATERAL/ENGLISH_REPOSITORY/RPA_03-2006_INGLES_0.PDF

Mueller, S.A., Anderson, J.E. and Wallington, T.J. (2011) Impact of Biofuel Production and Other Supply and Demand Factors on Food Price Increases in 2008. *Biomass and Bioenergy*, 35, 1623–32.

Nass, L.L., Arraes Pereira, P.A. and Ellis, D. (2007) Biofuels in Brazil: An overview. *Crop Science*, 47, 2228–37.

Natanelov, V., McKenzie, A.M. and van Huylenbroeck, G. (2013) Crude Oil-Corn-Ethanol – Nexus: A contextual approach. *Energy Policy*, 63, 504–13.

National Research Council (2011) *Renewable Fuel Standard: Potential economic and environmental effects of U.S. biofuel policy*. Washington, DC: The National Academies Press.

Nazlioglu, S., Erdem, C. and Soytas, U. (2013) Volatility Spillover Between Oil and Agricultural Commodity Markets. *Energy Economics*, 36, 658–665.

Nuffield Council on Bioethics (2011) *Biofuels: Ethical issues*. Available at: http://www.nuffieldbioethics.org/biofuels-0

Oberthür, S. and Pallemaerts, M. (eds.) (2010) *The New Climate Policies of the European Union*. Brussels: VUB Press (Brussels University Press).

Oberthür, S. and Pallemaerts, M. (2010) The EU's Internal and External Climate Policies: An historical overview. In Oberthür, S. and Pallemaerts, M. (eds.) *The New Climate Policies of the European Union*. Brussels: VUB Press (Brussels University Press), 27–63.

OECD (2008a) *Rising Agricultural Prices: Causes, consequences and responses*. Policy Brief. Paris: Organisation for Economic Cooperation and Development. Available at: http://78.41.128.130/dataoecd/1/36/41227216.pdf

OECD (2008b) *Biofuels Support Policies; An economic assessment*. Paris: Organisation for Economic Cooperation and Development.

O'Hare, M. [and 12 others] (2011) Comment on 'Indirect Land Use Change for Biofuels: Testing predictions and improving analytical methodologies' by Kim and Dale: Statistical reliability and the definition of the indirect land use change (iLUC) issue. *Biomass and Bioenergy*, 35, 4485–4487.

Oliver, E. (2013) *EU Biofuels Investment Development: Impact of an uncertain policy environment*. Agra CEAS Consulting Special Study. Available at: http://www.ceasc.com/Images/Content/Agra%20CEAS%20study%20on%20EU%20Biofuels%20Investment%20Development%20-%20Impact%20of%20an%20Uncertain%20Policy%20Environment%20-%20December%202013.pdf

Ott, H. (2014) Volatility in Cereal Prices: Intra- versus inter-annual volatility. *Journal of Agricultural Economics*, 65(3), 557–578.

Overmars, K.P., Stehfest. E., Ros, J.P.M. and Gerdien Prins, A. (2011) Indirect Land Use Change Emissions Related to EU Biofuels Consumption: An analysis based on historical data. *Environmental Science and Policy*, 14, 248–257.

Özdemir, E.D., Härdtlein, M. and Eltrop, L. (2009) Land Substitution Effects of Biofuel Side Product and Implications on the Land Area Requirement for EU 2020 Biofuel targets. *Energy Policy*, 37, 2986–2996.

Painter, M. and Pierre, J. (2005) Unpacking Policy Capacity: Issues and themes. In Painter, M. and Pierre, J. (eds.) *Challenges to State Policy Capacity: Global trends and comparative perspectives*. Basingstoke: Palgrave, 1–18.

Palmer, J.R. (2014) Biofuels and the Politics of Land-Use Change: Tracing the interactions of discourse and place in European policy making. *Environment and Planning A*, 46(2), 337–352.

Pangea (2012) *Who's Fooling Whom?: The real drivers behind the 2010/11 food crisis in Sub-Saharan Africa*. Available at: http://www.pangealink.org/wp-content/uploads/2012/09/PANGEA_food-prices_final.pdf

Parsons, W. (2004) Not Just Steering but Weaving: Relevant knowledge and the craft of building policy capacity and coherence. *Australian Journal of Public Administration*, 63, 43–57.

Patashnik, E.M. (2008) *Reforms at Risk*. Princeton: Princeton University Press.

Plevin, R.J., O'Hare, M., Jones, A.D., Torn, M.S. and Gibbs, H.K. (2010) Greenhouse Gas Emissions from Biofuels' Indirect Land Use Change Are Uncertain but May Be Much Greater Than Previously Estimated. *Environmental Science and Technology*, 44(21), 8015–8021.

Pollack. M.A. and Shaffer G.C. (2009) *When Cooperation Fails: The international law and politics of genetically modified foods*. Oxford: Oxford University Press.

Ponte, S. (2013) 'Roundtabling' Sustainability: Lessons from the biofuel industry. *Geoforum*. Article in Press, http://dx.doi.org/10.1016/j.geoforum.2013.07.008.

Pousa, G.P.A.G., Santos, A.L.F. and Suarez, P.A.Z. (2007) History and Policy of Biodiesel in Brazil. *Energy Policy*, 35, 5393–5398.

Puerto Rico, J.A., Mercedes, S.S.P. and Sauer, I.L. (2010) Genesis and Consolidation of the Brazilian Bioethanol [sic]: A review of policies and incentive mechanisms. *Renewable and Sustainable Energy Reviews*, 14, 1874–1887.

Puppim de Oliveira, J.A. (2002) The Policymaking Process for Creating Competitive Assets for the Use of Biomass Energy: The Brazilian alcohol programme. *Renewable and Sustainable Energy Reviews*, 6, 129–140.

Putnam, R. (1988) Diplomacy and Domestic Politics: The logic of two-level games. *International Organization*, 42, 427–460.

Qiu, C., Colson, G., Escalante, C. and Wetzstein, M. (2012) Considering Macroeconomic Indicators in the Food Before Fuel Nexus. *Energy Economics*, 34(6), 2021–2028.

Rajagopal, D. and Zilberman, D. (2007) *Review of Environmental, Economic, and Policy Aspects of Biofuels*. Policy Research Working Paper 4341. Washington, DC: The World Bank.

Rathmann, R., Szklo, A. and Schaeffer, R. (2012) Targets and Results of the Brazilian Biodiesel Incentive Program – Has it reached the promised land? *Applied Energy*, 97, 91–100.

Ribeiro, B.E. (2013) Beyond Commonplace Biofuels: Social aspects of ethanol. *Energy Policy*, 57, 355–362.

Richardson, J. (2012) New Governance or Old Governance? A policy style perspective. In D. Levi-Faur (ed.) *The Oxford Handbook of Governance*. Oxford: Oxford University Press, 312–324.

Roache, S.K. (2010) *What Explains the Rise in Food Price Volatility?* IMF Working Paper WP/10/129, International Monetary Fund. Available at: http://chede.org/chede/wp-content/uploads/2010/06/IMF-Reasons-for-Food-Price-volatility.pdf

Rohter, L. (2010) *Brazil on the Rise*. New York: Palgrave Macmillan.

Rosillo-Calle, F. and Cortez, L.A.B. (1998) Towards Proálcool II – A review of the Brazilian bioethanol programme. *Biomass and Bioenergy*, 14(2), 115–124.

Rosillo-Calle, F. and Johnson, F.X. (2010) *Food Versus Fuel: An informed introduction to biofuels*. London: Zed Books.

Sabatier, P. (ed.) (1999) *Theories of the Policy Process*. Boulder: Westview Press.

Sabatier, P. and Jenkins-Smith, H. (1993) *Policy Change and Learning*. Boulder: Westview Press.

Savvanidou, E., Zervas, E. and Tsagarakis, K. (2010) Public Acceptance of Biofuels. *Energy Policy*, 38, 3482–3488.

Sbragia, A.M. (2005) Institution-Building from Below and Above: The European Community in global environmental politics. In Jordan, A. (ed.) *Environmental Policy in the European Union: Actors, institutions and processes*, 2nd ed. London: Earthscan, 201–224.

Schaffel, S., Herrera, S., Obermaier, M. and Lèbre La Rovere, E. (2012) Can Family Farmers Benefit from Biofuel Sustainability Standards? Evidence from the Brazilian social fuel certificate. *Biofuels*, 3(6), 725–736.

Schaffel, S. and Lèbre La Rovere, E. (2010) The Quest for Eco-Social Efficiency in Biofuels Production in Brazil. *Journal of Cleaner Production*, 18, 1663–1670.

Schmitz, P.M. (2012) *Determination Basis for the Level and Volatility of Agricultural Commodity Prices in International Markets – Implications for world nutrition and policy formulation.* Preliminary Study. Available at: http://www.biokraftstoffverband.de/index.php/hunger.html?file=tl_files/download/Stellungnahmen_und_Studien/12-02-23-Ufop%20VDB-Schmitz_Vorstudie_ENG.pdf

Schmitz, T.G., Seale, J.L. Jr and Buzzanell, P.J. (2002) Brazil's Domination of the World Sugar Market. In Schmitz, A., Spreen, T.H., Messina, W.A. Jr and Moss, C.B. (eds.) *Sugar and Related Sweetener Markets: International perspectives.* Wallingford: CABI Publishing.

Schneider, A.L. and Ingram, H. (1997) *Policy Design for Democracy.* Kansas City: University of Kansas Press.

Schnepf, R. (2006) *European Union Biofuels Policy and Agriculture: An overview.* Congressional Research Service report RS22404, The Library of Congress.

Searchinger, T., Heimlich, R., Houghton, R., Dong, F., Elobeid, A., Fabiosa, J., Tokgoz, S., Hayes, D. and Yu T.-H. (2008) Use of US Croplands for Biofuels Increases Greenhouse Gases Through Emissions from Land Use Change. *Science*, 319, 1238–1240.

Sen, A. (1981) *Poverty and Famines: An essay on entitlement and deprivation.* Oxford: Oxford University Press/Clarendon Press.

Sengers, F., Raven, R.P.J.M. and van Venrooij, A. (2010) From Riches to Rags: Biofuels, media discourses, and resistance to sustainable energy technologies. *Energy Policy*, 38, 5013–5027.

Serafim, I. (2006) Brazil: Brazilian Biofuels Industry. US Commercial Service Report ID 137297. Available at: http://www.ic.gc.ca/scdt/bizmap/interface2.nsf/vDownload/ISA_5056/$file/X_8027507.PDF

Sercovich, F.C. (1986) The Political Economy of Biomass in Brazil – The Case of Ethanol. In Jacobsson, S., Jamison, A. and Rothman, H. (eds.) *The Biotechnological Challenge.* Cambridge: Cambridge University Press.

Serra, T. (2011) Volatility Spillovers Between Food and Energy Markets: A semiparametric approach. *Energy Economics*, 33, 1155–1164.

Serra, T., Zilberman, D. and Gil, J. (2011) Price Volatility in Ethanol Markets. *European Review of Agricultural Economics*, 38(2), 259–280.

Sheales, T., Gordon, S., Hafi, A. and Toyne, C. (1999) *Sugar: International policies affecting market expansion.* ABARE Research Report 99.14, Canberra: ABARE.

Shortall, O.K. (2013) 'Marginal Land' for Energy Crops: Exploring definitions and embedded assumptions. *Energy Policy*, 62, 19–27.

Sissine, F. (2007) *Energy Independence and Security Act of 2007: A summary of major provisions.* Congressional Research Service report RL34294.

Skjølsvold, T.M. (2012) Curb Your Enthusiasm: On media communication of bioenergy and the role of the news media in technology diffusion. *Environmental Communication*, 6(4), 512–531.

Skovgaard, J. (2013) The Limits of Entrapment: The negotiations on EU reduction targets, 2007–11. *Journal of Common Market Studies*, 51(6), 1141–1157.

Soares Ferreira, V. da R. and Passador, C.S. (2010) Potentials and Limits to Generate Employment and Income by the National Programme for Production and

Use of Biodiesel. *Organizações Rurais & Agroindustriais, Lavras*, 12(1), 20–33 (in English).

Söderberg, C. and Eckerberg, K. (2013) Rising Policy Conflicts in Europe over Bioenergy and Forestry. *Forest Policy and Economics*, 33, 112–119.

Soetaert, W. and Vandamme, E.J. (eds.) (2009) *Biofuels*. Chichester: Wiley.

Stattman, S.L., Hospes, O. and Mol, A.P.J. (2013) Governing Biofuels in Brazil: A comparison of ethanol and biodiesel policies. *Energy Policy*, 61, 22–30.

Stewart, I.M.T. (1979) *Reasoning and Method in Economics*. London: McGraw Hill.

Stewart, J. (2009) *Public Policy Values*. Basingstoke: Palgrave.

Streeck, W. and Thelen, K. (2005) Introduction. In Streeck, W. and Thelen, K. (eds.) *Beyond Continuity: Institutional change in advanced political economies*. Oxford: Oxford University Press, 1–39.

Swinbank, A. (2009a) *EU Support for Biofuels and Bioenergy Environmental Sustainability Criteria, and Trade Policy*. International Centre for Trade and Sustainable Development (ICTSD) Global Platform on Climate Change, Trade Policies and Sustainable Energy Issue Paper No. 17. Available at: http://ictsd. org/i/publications/50270/?view=document

Swinbank, A. (2009b) Presidential Address: EU Policies on Bioenergy and the Potential Clash with the WTO. *Journal of Agricultural Economics*, 60, 485–503.

Swinbank, A. and Daugbjerg, C. (2013) Improving EU Biofuels Policy? Greenhouse gas emissions, policy efficiency, and WTO compatibility. *Journal of World Trade*, 47, 813–834.

Swinnen, J. (2011) The Right Price of Food. *Development Policy Review*, 29(6), 667–88. An earlier version of this research is available freely at: http://www. econstor.eu/bitstream/10419/74923/1/dp259.pdf

Switzer, S. (2007) International Trade Law and the Environment: Designing a legal framework to curtail the import of unsustainably produced biofuels. *Journal of International Trade Law and Policy*, 6, 30–44.

Switzer, S. and McMahon, J. (2010) EU Biofuels Policy – Raising the question of WTO compatibility. International and comparative law quarterly, 60(3), 713–736.

Taheripour, F., Hertel, T.W., Tyner, W.E., Beckman, J.F. and Birur, D.K. (2010) Biofuels and Their By-Products: Global economic and environmental implications. *Biomass and Bioenergy*, 34, 278–289.

Talamini, E., Caldarelli, C.E., Wubben, E.F.M. and Dewes, H. (2012) The Composition and Impact of Stakeholders' Agendas on US Ethanol Production. *Energy Policy*, 50, 647–658.

Tarasofsky, R. (2008) Heating Up International Trade Law: Challenges and opportunities posed by efforts to combat climate change. *Carbon and Climate Law Review*, 2, 7–17.

Taylor, P.G. [and 7 others] (2014) Palm Oil Wastewater Methane Emissions and Bioenergy Potential. *Nature Climate Change*, 4, 151–152.

Teixeira de Sousa, P. Jr, Dall'Oglio, E.L., Marta, J.M. Sato, M., Brito de Azevedo, R.A. and Spindola, C. (2008) The Ethanol and Biodiesel Programmes in Brazil. In Mytelka, L.K. and Boyle, G. (eds.) *Making Choices About Hydrogen: Transport issues for developing countries*. Tokyo: UN University Press; and Ottawa: International Development Research Centre, pp. 118–40. http://www.idrc.ca/openebooks/413-0/

Thacher, D. and Rein, M. (2004) Managing Value Conflict in Public Policy. *Governance*, 17, 457–486.

Thompson, W. (2010) Biofuel Effects on Markets and Indirect Effects on Land Use and Food. *Journal of International Agricultural Trade and Development*, 6(1), 117–131.

Trostle, R., Marti, D., Rosen, S. and Westcott, P. (2011) *Why Have Food Commodity Prices Risen Again?* Report WRS-1103, United States Department of Agriculture Economic Research Service. Available at: http://www.ers.usda.gov/media/126752/wrs1103.pdf

Turba, R. and Fajardo, E. (2008) *Biofuels in Brazil*. Worcester Polytechnic Institute Electronic Projects Collection, E-Project-100608-221212. Available at: http://www.wpi.edu/Pubs/E-project/Available/E-project-100608-221212/

UNEP (2009) Towards Sustainable Production and Use of Resources: Assessing Biofuels. Paris: United Nations Environment Programme. Available at: http://www.unep.org/resourcepanel/Publications/AreasofAssessment/LandandSoils/AssessingBiofuels/tabid/56055/Default.aspx

US Department of Energy (2010) *US Biofuels Industry: Mind the gap*. Department of Energy: The Office of the Biomass Program.

US Department of Energy (2013) *Database for State Incentives for Renewables and Energy (DSIRE)*. Available at: http://www.dsireusa.org/ (accessed 1 December 2013).

US Geological Survey (2010) *The U.S. Geological Survey Energy Resources Program: U.S. geological survey fact sheet 2010–3100*. Available at: http://pubs.usgs.gov/fs/2010/3100/ (accessed 1 December 2013).

US Senate and House of Representatives (2005) *Energy Policy Act*. Public Law 109–58, 109th Congress.

US Senate and House of Representatives. (2007) *Energy Independence and Security Act*. Public Law 110–140, 110th Congress.

Valdes, C (2007) Ethanol Demand Driving the Expansion of Brazil's Sugar Industry. Special Article. *Sugar and Sweeteners Outlook*, SSS–249, 31–38. Washington, DC: Economic Research Service, US Department of Agriculture.

van Dam, J., Junginger, M., Faaij, A., Jürgens, I., Best, G. and Fritsche, U. (2008) Overview of Recent Developments in Sustainable Biomass Certification. *Biomass and Bioenergy*, 32, 749–780.

van Thuijl, E. and Deurwaarder, E.P. (2006) *European Biofuel Policies in Retrospect*. Report ECN-C-06-016. Petten: Energy Research Centre of the Netherlands.

Vis, M.W., Vos, J. and van den Berg, D. (2008) *Sustainability Criteria & Certification Systems for Biomass Production: Final Report*. For DG-TREN, European Commission. Available at: http://ec.europa.eu/energy/renewables/studies/doc/bioenergy/2008_sustainability_criteria_and_certification_systems.pdf

Vogler, J. (2011) The European Union as a Global Environmental Policy Actor. In Wurzel, R. and Connelly, J. (eds.) *The European Union as a Leader in International Climate Change Politics*. UACES Contemporary European Studies Series. London: Routledge for the University Association for Contemporary European Studies, 21–37.

von Witzke, H. and Noleppa, S. (2011) *The Economics of Rumpelstiltskin. Why speculation is not the prime cause of high and volatile international agricultural commodity prices: An economic analysis of the 2007–08 price spike*. HFFA Working

Paper. Available at: http://worldagforum.com/files/von_Witzke_and_Noleppa_
-_The_economics_of_Rumpelstitzken.pdf

Wallington, T.J., Anderson, J.E., Mueller, S.A., Kolinski Morris, E., Winkler, S.L., Ginder, J.M. and Nielsen, O.J. (2012) Corn Ethanol Production, Food Exports, and Indirect Land Use Change. *Environmental Science and Technology*, 46(11), 6379–6384.

Wilkinson, J. and Herrera, S. (2010) Biofuels in Brazil: Debates and impacts. *The Journal of Peasant Studies*, 37(4), 749–768.

Wind, I. (2010) *HS Codes and the Renewable Energy Sector*. International Centre for Trade and Sustainable Development (ICTSD) Programme on Trade and Environment Paper. Available at: http://ictsd.org/downloads/2010/01/hs-codes-and-the-renewable-energy-sector.pdf

Witcover, J., Yeh, S. and Sperling, D. (2013) Policy Options to Address Global Land Use Change from Biofuels. *Energy Policy*, 56, 63–74.

Wolbring, G. and Noga, J. (2013) Greening and Energy Issues: An analysis of four Canadian newspapers. *Journal of Sustainable Development*, 6(7), 88–112.

Wolf, K.D. (2008) Emerging Patterns of Global Governance: The new interplay between the state, business and civil society. In Scherer, A.G. and Palazzo, G. (eds.) *Handbook of Research on Global Corporate Citizenship*. Cheltenham: Edward Elgar, 225–248.

Wurzel, R. and Connelly, J. (eds.) (2011) *The European Union as a Leader in International Climate Change Politics*. UACES Contemporary European Studies Series. London: Routledge for the University Association for Contemporary European Studies.

Yano, Y., Blandford, D. and Surry, Y.R. (2012) From Ethanol Shuffle to Ethanol Tourism – Why the RFS does not make sense. *Choices*, 27(4), 1–4. Available at: http://www.choicesmagazine.org/magazine/pdf/cmsarticle_264.pdf

Zahariadis, N. (2003) *Ambiguity and Choice in Public Policy: Political manipulation in democratic societies*. Washington, DC: Georgetown University Press.

Zhang, W., Yu, E.A., Rozelle, S., Yang, J. and Msangi, S. (2013) The Impact of Biofuel Growth on Agriculture: Why is the range of estimates so wide? *Food Policy*, 38, 227–239.

Zilberman, D., Hochman, G., Rajagopal, D., Sexton, S. and Timilsina, G. (2013) The Impact of Biofuels on Commodity Food Prices: Assessment of findings. *American Journal of Agricultural Economics*, 95, 275–281.

Index

Printed and bound by CPI Group (UK) Ltd, Croydon, CR0 4YY